THINK!NG SAFETY

TACKLING COGNITIVE RISKS IN THE MARITIME INDUSTRY

CAPT. ASHOK MENON

INDIA · SINGAPORE · MALAYSIA

Notion Press

Old No. 38, New No. 6
McNichols Road, Chetpet
Chennai - 600 031

First Published by Notion Press 2019
Copyright © Capt. Ashok Menon 2019
All Rights Reserved.

ISBN 978-1-64650-872-3

This book has been published with all efforts taken to make the material error-free after the consent of the author. However, the author and the publisher do not assume and hereby disclaim any liability to any party for any loss, damage, or disruption caused by errors or omissions, whether such errors or omissions result from negligence, accident, or any other cause.

While every effort has been made to avoid any mistake or omission, this publication is being sold on the condition and understanding that neither the author nor the publishers or printers would be liable in any manner to any person by reason of any mistake or omission in this publication or for any action taken or omitted to be taken or advice rendered or accepted on the basis of this work. For any defect in printing or binding the publishers will be liable only to replace the defective copy by another copy of this work then available.

Dedication

This book is humbly dedicated to the families of all seafarers around the world, who silently supports the fraternity of seafarers with their thoughts, prayers and with the deepest concern of their safety at all times, as they brave the high seas to sustain the quality of life of peoples across the globe.

Contents

Preface   7

PART 1 Safety on a Leash

Chapter 1 Introduction   13

Chapter 2 Maritime Safety & Risk – Present Scenario   19

Chapter 3 Are We Doing It Right?   26

Chapter 4 Human Element – The Way We Are   36

Chapter 5 Is BBS a Panacea?   49

PART 2 I Think, Therefore I am

Chapter 6 A Peek Into the Human Brain   57

Chapter 7 Stress, Anxiety and Fatigue   85

Chapter 8 Distortions of Our Reality   98

Chapter 9 Focusing the Brain   131

PART 3 Hitting the Nail on the Head

Chapter 10 A Holistic Safety Strategy   151

Chapter 11 New Training for Cognitive Risk Awareness   170

Notes & References   190

Preface

A small airline with 3 passengers crashes into a remote area of Brazil and this news is picked up by the international news networks almost immediately and is broadcast across the globe by various news media. If the flight involved slightly larger numbers of passengers, then it was sure to hog a large segment of the news coverage and analysed in depth. Interest in airline accident stories are of great interest to people and hence widely reported.

Now consider this. A cargo ship with a full load of iron ore breaks up into two and sinks off the Australian coast in heavy weather and all hands are lost as she sinks. This news is only heard in the offices of shipping companies and organizations, but not prominent in the mainstream media. The loss of ships and seafarers are not as newsworthy as we might expect. A passenger ship accident gets a better response and interest, but cargo ships fail to get the spotlight in the international news media. Seafarers remain silent ghosts patrolling the seas and distributing the economic goods across the world, but with no acknowledgement of their sacrifices and no public interest in the lives that they lead.

Because many marine accidents involving cargo ships are seldom reported in the mainstream media, it does give the impression that the sea is a safer place and accidents are a rare sight compared to the airline industry. For those who track marine based accidents, it comes as no surprise that on any given day, somewhere in some part of the world, some ship experiences a collision, has an explosion, or is broken up in bad weather or catches fire with fatalities which go sadly unreported. This indifference to shipping accidents by the news media is not surprising as the industry seems to be one without the glamour of its sister industry, the airline industry. In shipping, it appears that no news is good news. Unless one reads marine industry focussed newsletters or blogs, one never comes to know the large number of casualties or losses that this industry is regularly facing. Unless of course, the cargo ship has had an accident after which some pristine coastal beach got dirtied due to the resultant oil pollution. That is sure to get eyeballs and interest from various environmental bodies.

Also, shipping is a very complex industry which taxes even an expert to understand its very own nature. For example, a ship owner from USA will have his ship registered in Marshall Islands, with crew supplied by an agency in Cyprus who uses seafarers selected from half a dozen different nationalities, technically managed from an office in Singapore, in which many different nationalities work. The complex nature of the shipping industry makes reporting of shipping news events a matter for domain experts only. The mainstream media prefers to stay clear of this.

The human losses and injuries are as tragic as those that happen in any other industries. So, if we must address the causes of these losses and injuries, we must look at a paradigm shift in thinking about safety and address more deeper issues to ensure safety training bites the real bullet!

However, it goes without saying that shipping is a very high-risk industry and no shipping adventure is successfully completed without the fluctuating levels of risk always ever ready to strike when the time is opportune, and the ship staff is caught off guard. There is no way that cargoes can be carried safely without living through such risks, but these risks should be watched carefully to prevent it from rearing its ugly head and causing accidents. Risk is our dancing partner all the way and we should learn to step correctly lest it trips us.

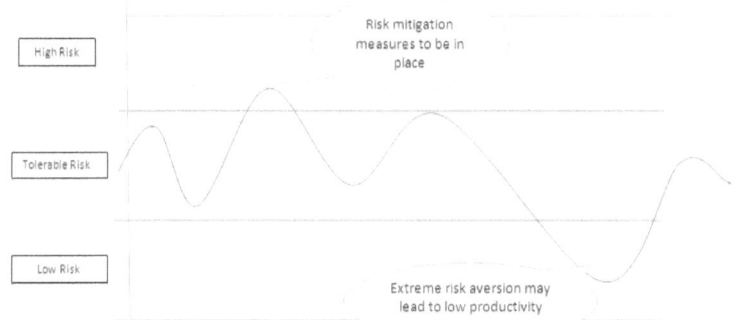

A few years ago, I came across the famous illusionary experiment known as the "Monkey Business Illusion." This was created by Daniel Simons, Professor in the Department of Psychology and the Beckman Institute for Advanced Science and Technology at the University of Illinois. He is the author of the book "The Invisible Gorilla." I was fascinated at how a patent scene unravelling in front of an audience was not perceived by many of the viewers and that got me excited when I connected this with safety issues on board vessels. We never talked about it and yet, we thought that by following all the rules, checklists and procedures, we had covered it all. Later, after delving into the fascinating new world of neuroscience, I realized that we need to wake up and see this inner world of deception that may be the real cause of many accidents and for which we may have ascribed some other wrong causes.

In many ensuing accident investigations detailing the causes of an accident, the term "human error" was meant to be a full stop after which we could not identify any deeper causes. The fact that 80% of all accidents are caused by human error was still not enough to make us delve deeper into the world of human fallacies! So, these new findings about that brain's myriad workings encouraged me to excitedly investigate these new topics of interest.

We may have heard that when we get a physical symptom of any disease in our body, it may have arisen long after some other deeper imbalance, probably of energy flows, manifested and was not corrected. This was akin to what I am trying to bring out, namely that our unsafe behaviours are a symptom of a deeper cognitive malaise to which we never give a thought or train our workers to recognize and prevent the onset of such behaviours.

This book aims to direct the reader to delve into newer areas of concerns in our understanding of safety. For more than 25 years, we have used the International Safety Management Code and Safety Management Systems to regimentalize the approach to safety, mainly by use of checklists and permits prior to any critical operations. This has vastly reduced the number of incidents, however, can this approach be continued for ever?

The reader should specially note that the information provided regarding the human brain and its functions is for informational purposes only and does not constitute medical advice, diagnosis, or treatment. You should consult with a qualified health care professional before making changes to your lifestyle, starting any new health care regimen, or if you have, or suspect you might have, a health problem.

PART 1
Safety on a Leash

Chapter 1
Introduction

Safety is as simple as ABC… Always Be Careful.

Picture this! You wake up from deep sleep in a hospital bed and you find yourself staring at the ceiling. One of your legs is fully plastered after having suffered multiple fractures. You are suddenly aware of a deep throbbing pain in your head. Then you realize that there are wads of bandage around your skull that is covering a bad head injury. Obviously, this is not what you set out to do on that fateful day when you climbed up an unsecured ladder in a ship's hold. You merely wanted to complete the mandatory inspections of the top areas of the cargo hold. However, the ladder, not being secured properly slipped and threw you down violently onto the hard deck below.

Sadly, that was all that you could remember of that steep fall. When you came to your senses, you saw the now familiar ceiling and also felt the sharp stabs of pains. If only you could go back in time and be put in the same situation as before climbing that fateful ladder! Surely, you would secure it properly as you had always been trained during the many training sessions your company had arranged. You would also now understand the real importance of these formerly "boring" safety sessions. Because, now you know! Safety saves!

Nobody likes an accident! And most ship operators and seafarers assume naively that these cannot happen on their vessels, and they are somehow always immune to it. But the sad fact comes home when a major loss or injury occurs on one of their vessels and the realization that we cannot undo it. Alas, the loss or damage is here to stay! And, an operator that suffered the incident will resign to the new reality and then perforce initiate the mandatory investigations as required by good industry practices. They set out to complete all the so called corrective and preventive actions that will conclude on the "root causes" of the incident. Very importantly, this should satisfy the authorities and Oil majors. This, they believe can save their face and allow them to trade for another day. The insurance may pay up for most of the losses and the financial impact is greatly reduced, even though their reputation may have taken a hit. With time, the lessons and seriousness of the event are lost. The cycle is ripe to repeat itself. That is what we see in the daily world of shipping.

So where are we going wrong? As a "normal" responsible company, the lessons were disseminated across the fleet, learning workshops carried out at various company seminars, documentation completed, audit proofing done and then it is business as usual. However, with the passage of time, we get a new type of accident after which the same cycle is repeated. The standard process of corrective actions is time tested and sufficient to get the nod from the authorities that are. So, did they too get it wrong inasmuch as the bureaucratic processes they favoured won the day? And boy! A "good" safety manager worth his salt knows how to game it. But did we win the main objective of achieving real safety? Of not ever repeating similar incidents ever again? That is the moot question.

2.1 NUMBER AND SEVERITY

This section provides general information about the number of marine casualties and incidents and their severity.

Figure 1: Number of reported marine casualties and incidents

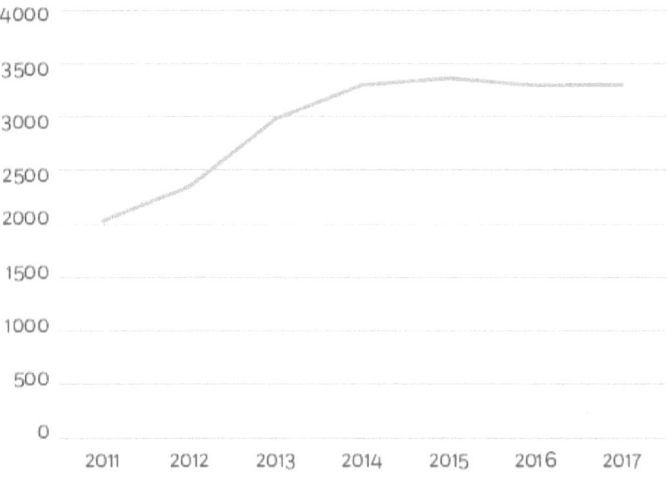

The total number of reported marine casualties and incidents is 20 616.
Since 2014, the number of reported casualties has stabilised at around 3 200 per year. However, comparisons with various sources suggest that under-reporting of marine casualties and incidents continues, with a total of 4 000 occurences per year being a best estimate.

Fig. 1.1: Extracted from Annual Overview of Marine Casualties and Accidents 2018 – EMSA

Our so-called safety trainings, being very scientific and logical aims to prevent accidents and unsafe behaviours. It has generally stood the test of time in making incidents and accidents an exception, rather than the rule. Even, the statistics of the recent years (see above figure 1.1) indicate that incidents and accidents are not increasing and seem to have plateaued. But the absolute numbers are still staggering and needs to be reduced drastically. Huge improvements are necessary to prevent the kind of numbers of accidents and losses that happen.

It is well known that humans are endowed with a powerful survival instinct that always guides him or her to avoid pain and injuries. But sometimes, that is not what we see in the real world and despite all the best safety trainings, we still have reckless behaviours leading to accidents for which we once again, use the same old rules to investigate and find solutions. We again come up with "sensible" sounding recommendations that are temporarily good for satisfying authorities and industry organizations, but sadly, leave the real root causes untouched. Or sometimes, it is believed that the real root causes cannot be influenced and corrected and therefore not practical! So why bother to look deeper?

The approach used for traditional safety trainings must be now closely reviewed to see if seafarers are learning to work absolutely safely or are still prone to accidental behaviours. In short, have we eliminated all known causes of accidents as it is only to be expected that they are now wise to these occurrences and can prevent them from occurring.

Listening to the common refrain from experienced seafarers and looking at safety statistics out of shipping companies, it can be argued that this has not been achieved and will not be achieved unless we fully understand why we are driven to unsafe behaviours and why many of us are programmed to take risks as a part of the human biological make up. What we cannot change may be managed better if we understand what is driving the inner core of our internal mechanisms and external behaviours.

New research findings in the field of neuroscience are out and indicate that humans are not just a mobile unit controlled from the skull box that decides what the rest of the body will or won't do! Also, each person comes with all shades of personality traits and learned behaviours that clearly shows that we can be different persons even if we sat through similar training sessions and we can still end up behaving with varying outcomes. Also, an individual itself can be a different person at different times complicating the dynamics of team formations. So, what is the real "we"? And what makes us so?

To get answers to these questions, let us go on a journey that will take us away from our usual trodden paths of safety trainings which was usually "knowledge" and competency based with case studies. We will also explore the labyrinths of the human mind and its frailties to explain why we humans cannot be always trusted to follow the linear & logical laws of safety.

Also let us try to understand how the human brain learns and how this learning is affected by various influences in the deeper realms of the brain.

In short, we look at how our "reality" is created around us and how we react to this perceived reality.

We may not find solutions for reaching the perfect score of "zero accidents and incidents." However, it is the author's view that without a deep understanding and awareness of the human inner complexities, we will not be closer to any solutions to create a safe nurturing world around us and others.

The author believes that even though we are programmed as a "human," we can still overcome these automatic regressions to unsafe tendencies by becoming aware of the way we interact with our environment and the perception of our realities that we create. Only then we can tame the unseen dangers of being influenced by what the commercial shipping world call as the "human element."

By understanding the capabilities and deeper vulnerabilities of human beings, we will have better success at arriving at solutions for our safety problems.

Until unmanned vessels are deployed on a large scale, the dependence on the human element will continue to remain as a central pursuit for a profitable and safe shipping industry. So, it makes eminent sense to explore how we have learned to think and react to the reality around us and lead us to a new awareness of the way our brain nudges our behaviours for good or for bad.

Chapter 2

Maritime Safety & Risk – Present Scenario

"Kelly was a seaman, and his life on the water followed a strict routine, which meant observing all the safety rules that had been written in the blood of less careful men."

– Author Tom Clancy, "Without Remorse"

Tomes have been written by most shipping operators to train their sea staff on how to operate their vessels safely. Ever since the advent of the International Safety Management (ISM) Code, shipping companies came out with their so-called Safety Management Systems (SMS). Ship operators and seafarers had to follow these stringently laid down procedures and checklists that tried to keep a check on arbitrary behaviours and unauthorized actions. They also served the purpose of educating seafarers on how the Company culture and values were to be preserved while carrying out various activities on board. Without doubt, the benefits of well written procedures and guidance have been enjoyed by many shipping companies as it reduced the number of accidents and losses.

Over the years, these procedures and checklists have become a necessary read for any seafarer joining a vessel to become fully familiar with all safe operations and regulations. Good monitoring and close follow-up by the prudent ship operators meant that they kept their accident and unsafe statistics to very low levels. However, many operators could not bring down these accidents and losses due to poor screening and training of their seafarers, apart from poor monitoring and support of the vessel operations. To add to the woes, operators in the pursuit of better efficiencies and safety, added more stricter procedures and checklists, thus burdening the same seafarers with more shipboard clerical work than they desired. Usually, such new instructions were introduced after a recurrence of a fateful incident from which new "lessons" were learnt. The new regulations for work and rest (Maritime Labour Convention 2006) that were introduced to improve safety and wellbeing of seafarers added to their problem, because of stricter working hours requirements. Now they had lesser working time to do their duties and complete all the documentation. Operators, being cost conscious did not want to add to their cost by offering extra pay for overtime work or add more manpower. But neither the operators nor the industry unions gave the seafarers any legitimate solutions to this conundrum. Operators did not like rest hour violations and were not pleased if any operation was stopped due to non-availability of rested personnel. A Catch –22 situation! They expected ship's personnel to be always ready and well rested for any Port State Inspections or Vetting Inspections, with no consideration for the crew who may have spent long tiring hours prior to arrival in port. The inconvenient truth of "fatigued seafarers" was not palatable and nobody found it convenient to discuss this subject. So, seafarers were left to their own devices and "adjusted" their rest hour records to show compliance to ensure commercial interests are taken care of. "Commercial pressure" is a dirty word and no shipping operator worth their salt will agree that they give this the highest priority while operating the ships. Their stated policy nowadays is that "Safety" is the highest priority and Master is given an "Over riding authority" to ensure that safety is supreme! However, is that always the real scenario?

To compound matters, during shipping downturns, shipping operators being very cost sensitive, tend to reduce manpower or essential spares and stores, forcing seafarers to compromise on safety. However, it should be put on record that good operators do exist who "walk the talk" and have seafarer and safety friendly policies. Generally, most of them are the good ship owning companies that have a much longer view of industry and their vessels. They take good care of their assets and personnel. And it must be added that many seafarers do make a very good attempt at following faithfully the well laid out procedures and regulatory requirements and take good pride in a ship well run.

So, let us look at the reality of a seafarer's life today?

The very dynamic nature of shipping makes the seafarer operate a vessel in such varying conditions that most other industries on the planet will not have to contend with.

Living in a metal box with unknown and undesirable effects of electro-magnetic radiations, their bodies exposed to ever changing time zones and likely hormonal disturbances, working with colleagues who are constantly being replaced at short intervals and coming from differing nationalities and cultural diversities, having to work with teams that are ever changing, operating the cargo at various international ports with varying geographical conditions, under constant pressure of deadlines and shipping profitability warnings, varying and sometimes extremes of environmental conditions, threats of hazards from the kind of cargoes carried and areas visited, sea piracy and armed robbery threats, being victims of high stake geo-political games, isolation and loneliness, nutritionally challenged foods that are mainly kept frozen for long periods of time, drinking fresh water that may be depleted of minerals, helplessness and frustration due inability to assist in family issues at home, constantly stressed about job security due to the nature of seafarer's short term job contracts, lack of healthy entertainment and sports activities, poor work and life balance, poor social life when back home, seafarer criminalization by authorities around the world, threat of abandonment, detention and jail terms at remote ports, lack of legitimate outlets to relax and travel in the ports visited, high risk and threat to personal safety are some of the serious issues faced by the seafaring community.

While efforts are being done to mitigate some of these issues, by and large, most are given only lip service and not tackled with seafarer's interests at heart. After all, the commercial interests are prime and that is what gets the focus of a shipping operator. This is a sad litany for the world of seafarers who are responsible for the transportation of 90% of the world's trade across continents. Their huge impact to keep the economies of the world well-oiled through their sacrifices is poorly recognised by rest of the world community. It is to the credit and great heart of our seafarers that they have silently taken on this responsibility to keep the world well fed and provided without expecting the gratitude of the world. The present scenario should be recognized by governments and shipping operators to give more benefits and attention to seafarers and ensure that regulations which make seafaring a friendly industry are encouraged so that more seafarers are attracted to it and come out in larger numbers for the good of the world economy.

It goes without saying that the main operator, the seafarer, is not a person we think can be compared to a usual industry worker ashore and who does not have to face the above conditions of existence. Sometimes, the joke goes that a seafarer is a "seaman" but not a "human"!

All governments and authorities should recognize the extraordinary sacrifices that they have to put up so that they are treated as important ambassadors of world trade instead of potential criminals and illegal immigrants who are coming to damage the destinations or countries they visit.

So how does all the above relate to safety of the people, vessel, cargo and environment?

If after knowing about all the above effects on a seafarer while going about his daily life is not enough, should we still expect that the seafarer to be a very safe and fully alert worker at all times and bring the ideal safety results for the organization as is expected in other shore-based industries? We have the procedures, checklists and instructions for him. However, will that alone be justified to expect full-hearted compliance in the way it is structured today?

Even though we cannot eliminate many of the deleterious effects of seafaring life, we can probably take some actions that may mitigate some of these effects. While most shipping companies know the answer, they feel that for an industry that goes under whenever recession rears its ugly head in the world economy, it is imprudent to spend more on seafarers as most of the daily running costs of a vessel are already due to them. Any further increase would make the venture unprofitable. Ideal actions that some seafarers demand are extra manning to reduce the work overload and fatigue, more work and life balance, a more safer environment, less criminalization of seafarers, more respect and recognition for their profession, good quality training, nutritionally better diets, better entertainment, physical exercise facilities, better contact with the outside world and access to news from home, social security and pensions, better medical care when on leave, family support, shorter tenures etc.

These will go a long way in giving them a more joyful work environment and pride in their lives.

Having said this, will these alone improve safety performance? No, as there are other dimensions to living in a hazardous environment such as a ship than just welfare and self-pride. In addition to what we call good safety training, we have to also empower the seafarers with new knowledge of what makes them behave as a "human." A "human" is not a robot who can be programmed to behave with similar results each time the same stimuli is experienced. A human always comes up with something new, sometimes to their detriment!

So, what are the main present-day expectations from seafarers by shipping companies to remain an ideal compliant worker on board?

- Ensure all necessary documentation are correctly recorded and filed.
- Never break any applicable regulation deliberately or accidentally.
- Absolute transparency and honesty.
- No conflicts with Charterers and commercial entities.
- Report all Non-conformities and compliance issues.
- Good planning and safe execution of duties.
- Good Safety culture and a culture of Risk assessment.
- Timely and proactive maintenance of vessel and equipment.
- Good people management and effective work culture on board.
- Ensure very high safety and occupational health onboard.
- Alert, well rested and competent teams for all activities.
- PSC/Vetting/Other inspections: Zero non-conformities/observations.
- Zero Marpol (Marine Pollution regulations) violations.
- Completes his/her employment contract as per agreement and without deviations.
- No Drug & Alcohol policy violations.

If the above conditions were met by seafarers, they would be the darling of the operators and generally would be most sought after for employment. However, to be fully satisfying the above conditions, it is the author's view that the training the seafarers have received so far is not complete to make them reach their ideal potential.

If we were to list the kind of trainings most seafarers undergo apart from their competency courses, it would generally look like this:

- Various regulatory knowledge trainings (Marpol, OPA 90, VGP, SIRE etc.)
- Safety seminars (Case studies, workshops etc.)
- Company seminars (Company performance, values, policies, commercial impact, safety performance etc.)
- STCW modular courses for Safety (Basic and Advanced & Refresher training)
- Safety equipment training (various)
- Operational equipment/machinery (Navigation, Engine room, Special tools etc.)
- Skill trainings (Ship handling, Engine room simulator, ECDIS, Pumpman, Fitter, Deckhand, Auditor)
- People skills (BTM, ERM, MRM)
- Commercial knowledge (Ship business, P & I/H & M workshops)

Even if the above courses cover the entire gamut of the trainings that a seafarer must undergo to remain certified to sail on board vessels, there are important aspects of human behaviour and performance, the understanding of which are sorely lacking today. Even though the concept of "human element" explains new factors that affect our safety decisions and actions, it is probably high time that companies devote more time and effort to disseminate these new thinking to proactively avoid future accidents and losses. These will be discussed in the next chapters.

CHAPTER 3

Are We Doing It Right?

"It will never happen to me" – Captain E. J. Smith (Captain of the Titanic)
— quoted in the press just before sailing

For more than 20 years, the shipping industry has sailed with the concept of Safety Management Systems (SMS), a risk-based system that strives to bring down accidents and loss statistics to a negligible level. The SMS generally came in few volumes of instructions and guidelines for every aspect of ship operation that dealt with people, processes, equipment and environment based on current regulatory requirements. These voluminous texts were to be mandatorily read and understood by every seafarer before taking up active duties. Every set of instruction was to be fully complied with and the resultant documentation filed away for proof of activity. Ideally, this was a foolproof method to standardize all activities and to prevent dependence on memory which can play truant at times.

With sustained efforts and continuous monitoring, some companies have reaped enormous benefits from these interventions. Many others did not get the same level of returns as the commitment and focus were not as sharp and sustained. If we were to look at what the everyday practice of doing safety onboard vessels is, we will see more or less a standard pattern in the outcomes across the shipping industry. To a large extent, the

Safety Management System of most companies are near clones of each other, give and take a few sets of procedures/checklists etc. Any seafarer switching from one company to another will not be "at sea" as similar processes and practices exist across various companies. This standardization of procedures and checklists may appear to be beneficial in terms of reducing the learning curve during this change of company, but this itself becomes the root cause of complacency and indifference preventing a closer study and evaluation of the new Company's procedures and compliance requirements.

Let us look at where we may be going wrong as we push the so called "good practices" with good intent, but sometimes with the likely effect of getting wrong outcomes.

From the earlier chapter, let us revisit some of the main expectations from seafarers by companies to remain an ideal compliant worker on board?

- **Ensure Company SMS is Fully Complied with.**

    Rules and procedures are designed to limit system variability. If they are followed, they may help to avert accidents up to a point, but they also prevent beneficially novel behaviour from emerging too. If the SMS procedures written down are too lengthy or ending up micro managing every operation, it won't be long before the instructions are disobeyed or missed due to the sheer stress of reading and complying with the voluminous procedures. Every company should look at keeping the set of procedures at just the right quantity and not get excited about making the longest and comprehensive set of procedures. This will reduce unnecessary compliance issues and also the stress of remembering unwanted set of procedures after which documentation (that most hated duty!) is to be completed. A very exhaustive SMS sows the seeds of its own destruction! Companies tend to forget that some seafarers are visual learners and introducing pictures and process diagrams into their procedures may get the right attention. In the coming days, it won't be a surprise to see companies come out with interactive and intelligent SMS (integrated with Artificial Intelligence and Virtual Reality) to make understanding, compliance and recording easier.

- **Ensure All Documentation are Correctly Recorded and Filed**

    The recording of documentation and the filing systems is usually taken very seriously by ship staff as they are aware that the performance of the vessel in an audit or during vetting inspections have a tremendous effect on its trading and earning potential. This need for extra seriousness leads to an attitude of "ensure the paperwork is done"! So an overzealous officer will resort to preparing many checklists and permits beforehand and filed away (with only the date and time to be inserted!) thus defeating the very purpose for which these procedures were designed. Of course, seafarers resort to these types of behaviors as they must have been caught at the end of the wrong stick when trying to comply honestly and had been punished. Any previous bad experience becomes a teacher par excellence! So, this extreme need to have documentation properly completed will need to be re-considered as newer forms of safe recordings (encrypted audio/video etc.) can also be introduced as evidence of compliance, saving time. Voice files can be digitally converted to text at an appropriate time. A digital log book or document is a sure way forward and this will save time and assist in better compliance. Of course, we must be wary of "fake" and "deep fake" digital media.

- **Never Break any Applicable Regulation Deliberately or Accidentally**

    While the deliberate breaking of any regulation should be recorded and reported, how many companies have encouraged their seafarers to report or record accidental breaking of regulation? For example, if a vessel accidently made a small oil pollution in a remote sea area and this was not caught or observed by any shore authorities, how many operators can confidently state that their staff are trained to report it to coastal authorities immediately without fear? And how many companies will voluntarily report this incident to the Flag state authorities? Some good operators may do this without a second thought, but most will shy away and thus pass on to their ship staff a long-lasting lesson in non-compliance. Organizational values are in question and is a powerful influencer.

- **No Conflicts with Charterers and Commercial Entities**

    How many Masters and senior officers can honestly state that they will operate their vessels with safety as the primary goal and not commercial expediency? This belief is strongly influenced by how the organizational leadership has communicated this very important priority. In various modes of communication, Companies give away their subtle fears and concerns of remaining profitable. Their styles of expressing fears of markets and commercial performance can sometimes be contagious and this leads to a fawning nature by the ship staff to obey any orders by the commercial entity. It won't be long before Masters will easily get into decision making that favour commercial interests without realizing that they have put the vessel's safety in jeopardy. There are umpteen cases of these happening and Companies have still not fully grasped this lesson.

- **Report All Non-conformities, Near Misses and Compliance Issues**

    Concepts of "No-Blame culture" or "Just culture" and the mandatory reporting of all non-conformities and observations have to a large extent brought about a culture of transparency and better communication. Many Companies are not amused when such Non-conformities or Near misses are repeated at regular intervals. This clearly indicates that the lessons are not being learnt and a preventive mindset is not being established. Also, the "documentation overload" is felt amidst the prevailing intense workload to safely operate the vessel. Non-conformities and reporting that was once seen as a good practice may be seen as a nuisance and an "unnecessary work" that is best avoided.

A "near miss" report is seen as another bureaucratic headache that needs to be "complied" with. Most companies have established minimum numbers of Near misses (NM) to be reported every month and after many years of reporting, the creative juice is drying up. Every Master has to plan how he or she can drum up the minimum number of NMs by the end of the month? Else, the Safety manager will be breathing down his/her back demanding the minimum numbers as that will demonstrate to external auditors and Vetting inspectors that the Company has a robust accident prevention system. This is the most ridiculous demand as the function of Near misses is not understood and it is expected that the more it is reported, greater the number of accidents has been prevented. Nothing could be farther from truth! The author believes that most Near misses (not all!) are reported just to complete the quota of NM reporting as a compliance matter. Masters encourage ship staff to report and guarantee no negative actions against them, but the sheer indifference and natural mistrust among staff prevents them from being truthful. Secondly, due to poor training, many seafarers have not clearly grasped the concept of this term. Hence, they do not recognize it when a near-miss occurs and it goes unreported.

However, the biggest concern is that the Safety departments in the ship operators' offices do not seem convinced about such reporting or they have themselves not delved into the advantages of such reporting. The biggest problem is that while NM reporting has been a regular practice for many years, no one is clear what a company's targets should be in the long term. Should NM reports be reducing, increasing or remaining same? Which one of these indicates a safer state of affairs? If NM reports are increasing, does it mean that we have a more aware and alert ship staff? Or does it mean that the staff is still showing lack of alertness in daily activities? If it is reducing, does it indicate that the vessel is safer? Or does it show reduced levels of alertness and non-observances of NMs? No one is sure how to interpret these results. If so, is it really worth measuring? The theory says that over a long period of time, when safety awareness is peaking, NMs should also fall. However, we still cling to the reporting of minimum numbers of NMs as a display of our commitment to safety reporting. How we continue to dance to the needs of auditors and inspectors!

- **Good Seamanship and a Culture of Risk Assessment**

  Long before ISM and SMS came into the seafarer's daily jargon, there were simple terms to ensure that a good job was done. In those days, this term was defined succinctly as "Good seamanship"! Today, we understand it as good risk assessment (RA), excellent planning, consistently safe practices and high levels of competency. Without doubt, a culture of Risk assessment is very much required and the author believes that this is one of the most important learnings that a seafarer will need to be equipped with. However, RA should not be "something" to be complied with. It should be the most creative exercise to discover all the varieties of hazards prevalent, however improbable, in the activities being undertaken. It must become a cultivated "mindset." Many accidents that happened in the past can be attributed to lack of creativity in thinking about hazards.

But are Companies giving this high importance? Are they checking to see what is taught and what is learnt in a classroom RA training? Or are they concerned with a piece of paper stating RA training is completed successfully? The author believe that this training will be one of the most critical learnings and operators should insist on ship staff ensuring that controls are in place to prevent all known and identified hazards from being activated.

- **Alert, Well Rested and Competent Teams for all Activities**

Ensuring high alertness and rested crews for various critical activities is usually achieved by good work planning using rest hour software. However, accidents happen when poorly rested crews are thrust into active duties especially during night hours and heeding unnecessary commercial pressures to complete the operations. Whereas many shipping operators have given clear instructions to Masters that he/she can use over riding authority to stop work if the crew is not properly rested, this is seldom practiced and not many Masters seem to have the gumption to execute it. Also, a clear support for the Master when he/she deploys this action is not forthcoming from many operators. Masters have to think very hard before sticking their necks out and stopping operations to give the much-needed rest to their tired crews. After doing this right action, they rarely get the appreciation for this proactive action. At best, no negative actions are taken against such Masters. How can such good practices be encouraged if the top down approach is rather negative? At the same time, Masters who have used this authority successfully in well run companies should be very careful not to over use this authority as it may become an easy route to non-performance and complacency. A very good and well considered decision should be taken when exercising the authority so that safety goals are achieved and the interrupted work activity should be resumed at the earliest opportunity once a risk assessment is made and found satisfactory.

- **Mentally and Physically Fit Crew Working as Productive Teams on Board:**

    Investing in the mental and physical health of the seafarers is never seen as an essential and important duty of the operator. Given the conditions under which seafarers live, it is surprising that seafarers successfully achieve the goals set out by their leadership and manage to deliver safe outcomes most of the times. A small investment in their mental and physical wellbeing will reward every operator with seafarers who can go that extra mile to achieve better targets with lesser hardships. Nurturing the seafarers is essential to keep them positively motivated and happy to continue working in the extremely trying environment they are exposed to. Some manning company staff will smirk at these statements by stating that the seafarers are already a pampered lot and it will get more difficult if we accede to their increasing demands. It may be true of some seafarers who will try to take undue advantage of the opportunities. However, let us not forget that a large majority are driven by their professionalism and pride in doing a good job. It is not suggested that that we handle them with kid gloves. Instead, a "listening" attitude can do wonders and improve not only performance, but also loyalty to the company. How many seafarers have left their employment in companies due to minor slights faced in company offices than due to major inconveniences faced on board the company vessels? The author has frequently heard senior managers of manning companies stating that some of their officers are faithful to the company because they had taken care of them when they had a special problem or family issue which was dealt n a compassionate manner. This actually is never forgotten by most seafarers. Even though relieving an officer midway through the contract may be an extra cost at that time, in the long run, the company will save when they have their own reliable faithful and competent staff serving their contracts as planned, instead of spending more money packing off untested persons after they create issues onboard.

To sum up, there are so many little things that we can do to get us a win-win situation as far as the operator-seafarer is concerned. First, a transparent organization that is willing to "listen" to seafarer plaints is the need of the hour. Experienced senior seafarers are always heard complaining that present day seafarer quality is not what it used to be as when they were sailing. If the best quality candidates must be brought to this profession, then it is a matter of paying the right price. After all, the labour market is highly competitive and the choices for an attractive career are multiplying in most seafarer supplying nations. If we can't get the best candidates, then we should make all efforts to give the selected ones a very high-quality training program to make them competent to the standards we require. Complaining and wringing our hands will take us nowhere.

Another important factor in attracting the right talent is the image of the company the seafarer wishes to serve. Sometimes, it is not about the money offered, but about the organizational values, beliefs and work culture where emphasis is on high levels of integrity and transparency. Companies such as this will find it easy to attract and retain the right people as their values resonates well with the new joiner and injects a "feel good" factor while serving such companies. Ensuring that companies raise their bar by "walking the talk" in matters of transparency and integrity than just for the sake of compliance, will go a long way in ensuring that the right persons are attracted to its work force.

As mentioned earlier, shipping is a very complex affair and not very easy to understand for the laymen. There are many levels of stakeholders with varying goals to achieve, some not aligned to each other. A crew manning agency may be recruiting for half a dozen companies, owned by different shipowners. These shipowners may be using 3rd party ship management companies to operate their vessels. If the high standards of the shipowner have to percolate down to the joining seafarer, the point of contact for a large number of seafarers being the manning agency, it is imperative that the manning agency staff is fully geared to instill the values and beliefs of the ship owning company. However, this is a tall order as crew manning office staff have their own work pressure and sometimes are not able to give time for such briefing. Or, they themselves are not familiar with the culture of the ship owners as they work with different ship owning entities. Some companies do have pre-departure briefings for the seafarers, but this can sometimes turn into a routine affair, done more for the compliance purpose than for influencing behaviour changes.

We can start by fixing these above-mentioned issues before we look at a more complex, but little understood term, now famously known as the "human element."

CHAPTER 4

Human Element – The Way We Are

"Concern for man and his fate must always form the chief interest of all technical endeavours."

–Albert Einstein

Many Shipping operators and owners are probably looking forward to the new emerging technologies that will bring about the advent of autonomous ships. Their main issues with human manpower have been cost of manpower and the unpredictable safety performance it entails. It is well known that a huge part of the daily running expense of a vessel is the manpower cost. All other costs cannot be further reduced as it will have a negative impact on the safe operations of the vessel. However, any savings that a manpower reduction can achieve will bring huge dividends to the ship operators bottom line.

The other main issue with using humans in the operations is the globally recognized "human error" problem. Unlike a computer, a human brain's complex workings have still not been fully deciphered by the scientific community to predict human behaviour accurately. The advancements in AI (Artificial Intelligence) technologies will probably make prediction of human behaviour come to fruition in the coming years. However, the day is not far away when each seafarer carrying out a critical duty will be wearing a lightweight head gear which will track his or her brainwaves, displaying brain states, attention and alertness data. This will be used to ensure that the right kind of brainwaves are present before critical duties are undertaken. And any brain activity that predict a potentially unsafe behaviour will be immediately flagged off and nipped in the bud.

The understanding of the construct of the brain and the functions of the various parts has only started emerging recently using new technologies like *f*MRI (functional MRI) which gave new insights into the workings of the brain. So, why would a shipping operator not desire to reduce human errors and eliminate huge losses, tangible and intangible, caused by accidents and injuries? These errors seem to recur with amazing regularity despite the best trainings that the industry can offer. Without doubt, this "human error" issues will be solved with the world changing over to vessels that will operate autonomously and provide huge savings in costs and reduction of safety incidents. But autonomous ships are still at a nascent stage of evolution and it will be a long while before it will be a common sight in the international trade arena, and that too on a large commercial scale. It may succeed initially in the coastal trade areas. And of course, it will be a great challenge for regulators to come up with the right set of regulations for autonomous ships that will allow the industry to sustain itself in the long run. The transition phase with traditionally manned ships and autonomous ships plying in the same sea lanes can also give rise to unknown challenges. However, the ship operators must continue to rely on the human brawn and brain to operate their vessels at sea for a long time to come. So, we can safely assume that the human error problem is here to stay. Or should it be so?

Industry has ascribed a name, "human element" to describe all those variables that come into an operation attributed to the human output while directly or indirectly carrying out work activities. Hence, the term human element, which is generally characterized with a negative connotation is seen more to do with the foibles of human nature than with the sum total of human expressions and behaviour. It sadly remains a nuanced expression to explain why things may go wrong because of the so called imperfect "human element" residing within each of us. It is this "human element" that we need to look and try to understand to mitigate such negative occurrences.

Without doubt, this same "human element" has not been given recognition for the countless numbers of positive human interventions that allowed an operation to succeed or prevent accidents. Good training and high standards in ship operations have given huge percentage of successes in shipping voyages and accidents remain an exception rather than the rule. However, for the purpose of these chapters, we will in general use the term "human element" to be those expressed characteristics of a human being from which emanates a proneness to error caused by non-standard reactions or lack of reactions to the multitudinous stimuli bombarding us from the external world.

Analysis of shipping disasters in recent years has shown that "human element" is of central importance in understanding the root causes of accidents and losses. The loss of life, the impact on company profitability and reputation, and the enormous environmental impacts that can result from the loss of a vessel remain a grave concern.

Many studies have brought rising awareness on human issues that are involved in many marine incidents. At the same time, the occurrence of marine incidents continues unabated.

It is now vital to make a clear connection between these human issues and the business success of those who make their living from the shipping industry – whether on ship or ashore. Specifically, everyone involved needs to understand that they, themselves, are the human element. From the board room to lower most ranks of the ship, everyone is in a veritable bear hug as they push their efforts for the success of their maritime adventures. They may appear to be disparate entities, but the action of each influences the action of others in this bear hug. Their continued business success depends on how far they can manage their own behaviour along with the behaviour of those around them.

Traditionally, we have approached all safety issues as a problem of "linear cause and effect" model by which we assume that we can finally cut out the cause at its root, the root cause analysis approach. When an activity is being carried out that later results in an incident, the person will never notice that the series of actions being taken are playing out as part of the linear "cause and effect" structure. It is only after an accident; we can obtain clarity of the events that made it seem to be an outcome of this cause and effect. In short, we can benefit from this knowledge only with hindsight. Only if it ends up as an accident, we strive to learn lessons. If the event had passed off without an accident, no one would have taken the effort to take stock and see if there was a possibility of accident happening. Also, no one seeks to learn any lessons from that work activity. In the real world, how many daily activities happen where accidents are narrowly missed, and no one bothers to re-visit the activity. That activity will pass off as another successful and uneventful one.

Even if a "human error" is actually made in a work activity that leads to a major accident, it is usually never that one person's "human element" that contributed 100% to the accident. Every person as part of a larger team, who is carrying out each action is making these decisions on the basis of the thought processes that are constantly playing out as undercurrents of the mind. Such thought processes are never independent of the world and situations around us. These are strongly influenced by outer influences that came in from company culture (how performance is defined by company in terms of efficiency or effectiveness, safety culture in practice, organizational values, onboard leadership influences) and personal background (personality traits, peer pressure, health status, past experiences and upbringing).

So, when we deem that "human error" occurred, we are looking at the universe of humans that directly or indirectly influence the running of the company and vessel, rather than one person who may have been the proximate cause of the accident. The person or team that directly caused the accident is only a small subset of that universe. Can we then ascribe blame on any person or teams for major accidents?

So, whenever a major accident happens, the knife is taken out to cut the root cause of the accident which was arrived at using the linear cause and effect principle. After that, the lessons learnt are promulgated across the fleet leading to ever-increasing rules and checklists. It is expected that in time, all seafarers will be able to carry out all operations with 100% certainty of safe completion and all possible errors would have been plugged for eternity to come.

However, in the real world of humans, the environment around us does not cater to this black and white outlook of the world. There are so many shades of grey because of which we need a better model to learn from. Experts have come out with the systems model to explain the complex and unpredictable nature of the world and relationships. It is out of these interactions that our behaviour – both good and bad – emerges.

In this view, it is recognized that designers of any safety procedures cannot visualize all situations and contingencies. Hence, the seafarer must be given some leeway to cope with the unexpected results.

Secondly, as any organization is a dynamic organic body, no fixed rules can always be applied to correct the situations that may go wrong, but new actions and results will tackle the emerging situations when the probability of things going wrong is realized. Any such changes must be quickly recognized and action to correct the adverse effects must be put in place without resort to the fixed rules laid down. This flexibility is necessary and can be successfully executed only if the whole organization recognizes the truth of systems effect and its interactions across the organization. For example, announcing exceptional achievements in safety can be good, but can also allow the seafarers to become complacent due to the pride inbuilt and the success promulgated. Safety is not something we achieved, but something we are always achieving.

So, let us recognize that we do not operate in a perfect world and it is normal for us to make mistakes. We do not operate in isolation and each of our actions are products of wider organizational factors and influences.

Today, many companies swear by a "Just culture" after having moved from the "No blame" culture.

A 'just culture' is founded on two principles, which apply simultaneously to *everyone* in the organization: (1)

- Human error is inevitable and the organizations' policies, processes and interfaces must be continually monitored and improved to accommodate those errors.

- Individuals should be accountable for their actions if they knowingly violate safety procedures or policies.

Achieving both these principles can be enormously challenging.

- The first principle requires a reporting system and culture that people can trust enough to make the necessary disclosures. Their trust develops out of the way the second principle is implemented,

- Specifically, from the way in which the organization defines, investigates and attributes accountability for whatever its staff disclose.

We also know that cultural influences affect outcomes, and this is something that cannot be wished away.

When we select a seafarer and put him or her onboard a vessel, it is akin to taking some water of a particular land well and mixing it with the ocean. It may appear to be indistinguishable as a water molecule, but it will still have the properties of the well it came from. The seafarer may behave as a seafarer should, but will unfailingly carry the cultural baggage and behaviours of the location from where he or she came from. These behaviours will bear a great influence on the final outcome of the group or teams that they interact with.

One such factor that strongly influences patterns of thinking and behaviour came from comprehensive and influential study by Geert Hofstede (1967–1973) which examined cultural differences in organizational and human resource management practice. (2)

He categorized these characteristics, which he called as "cultural dimensions" into four main ones: Power distance, Uncertainty avoidance, Individualism/Collectivism, Masculinity/Femininity. We can look at the relation between each dimension and its effect on the safety culture. (3)

The Relationship Between Power Distance and Safety Culture

Power Distance (PD), that is the extent to which members of the society accept that power in institutions and organizations is distributed unequally. When examined from organizational point of view PD is related to the degree of authority's centralization and autocratic leadership.

Employees coming from high PD:

- prefer managers' forming rules and regulations about safety beforehand, standardizing them and dictating themselves as commands complied with the hierarchical structure.

- employees think all the responsibilities about workplace safety belong to management and they have no other responsibility rather than obeying rules.

Managers who face with this kind of employee structure can determine a more efficient safety culture with such mechanisms as centralized structure, rigid rules and procedures, one-way top-down communication and strict supervision.

Employees coming from low PD:

- want to participate in every process of safety management, express their opinions and suggestions clearly, use initiatives when there is a safety problem in workplace

Thus, when worked with such a labour force, behaving in compliance with this is beneficial for managers.

The Relationship Between Uncertainty Avoidance and Safety Culture

Uncertainty Avoidance (UA) refers to the extent to which members of a particular culture feel anxious from uncertain and complex situations. In high UA cultures, individuals desire predictability and structuring in their organizations, institutions and relations. In order to cope with the anxiety resulting from uncertain situations; individuals feel the necessity of following rules and complying with behavioural codes defined with certain boundaries.

In Hofstede's study it is observed that in high UA cultures, individuals are subjected to more strict rules and laws and there are various safety measures to prevent uncertain or extraordinary situations. The higher UA gets, the less risk-taking tendencies about safety will be preferred by individuals. In other words, they will prefer the structuring of all systems, policies and rules without safety risks. In UA cultures employee expect their safety to be given importance by superiors.

In low UA cultures, individuals are subjected to fewer rules as possible and unpredictable risks and complexities are not very irritating. Low UA employees are less emotional, more tolerant towards uncertainty and willing to take risks. Thus, by taking employees' attributions in this type of culture into consideration, in managerial, safety applications, instead of regular basis safety training, a training program which are structured in frame of changing training needs and used interactive methods; safety management system that can be updated according to changing environment conditions with modern technological methods; empowering employee instead of over protecting them should be preferred in order to determine an efficient safety culture.

Relationship Between Individualism/Collectivism (I/C) and Safety Culture

Individualism shows the relations between the individual and group. Individualistic behaviours emphasize on personal benefits and employees are expected to take care of themselves. In these types of cultures individuals' having a successful social image is more significant than relationships and traditions.

In collectivistic cultures, individuals perceive themselves as a member of a society before identifying as individuals; group is the main factor determining beliefs and values. In these kinds of cultures, individuals grow up in extended families or socially cooperated groups; their loyalty to a group, tribe or village push them to protect their group's interests. Individuals' own opinions and beliefs do not differ from opinions and beliefs of the group they live in.

The explanations above demonstrate that I/C dimension is closely associated with "employees' involvement" and "risk perception" as dimensions of safety culture/climate. When examined from "employees' involvement" dimension, in high individualistic cultures it is more common independent thinking and taking initiative. If collectivism is a highly dominant, individuals avoid from expressing their personal opinions or visions when they have to face a critical decision-making situation. In obeying safety rules, instead of taking initiatives individually, following others and imitating them might prevail.

On the other hand, because individualism is related to talking about more direct communication and problems, it will benefit more in order to develop a positive safety culture.

I/C dimension is closely associated with risk perception in terms of safety. Individualism requires individuals to protect and put themselves priory. In individualistic societies individuals believe in personal success and importance of individual rights and struggle for them; it is expected to be responsible from oneself and their immediate family. In collectivistic cultures, individuals see themselves responsible from their extended families, close environments (relatives, friends etc.), groups, organizations and countries they are linked to more than their own responsibilities.

In economics and social psychology literature, distinction between collective and individual decision-making and its effect on risk behaviours have been focused on. For instance, a study by Shupp and Williams (2008) suggests that groups are more risk averse than individuals in high risk situations and group decisions show minor differences than individual decisions (Beugelsdidk and Frijns 2010). This shows the low tendency of risk-taking behaviour in collectivistic cultures. Chui et al. (2010) stressed that individualism can be linked overconfidence; in more individualistic societies, decisions are taken by individuals and these decisions are made on the basis of individual's feeling of security (Beugelsdijk and Frijns 2010). In collectivistic cultures, everyone expects an initiative from the other in order to change the unquestioned dependency.

In this context, in determining an efficient safety culture it is crucial for organizational management to behave by taking employees cultural backgrounds (I/C) into consideration. Towards individuals coming from individualistic societies, participative and empowering approach should be considered; more detailed personal safety trainings should be given; safety should be emphasized as a personal responsibility. While in collectivistic cultures, to generalize safety culture, safety can be shown as a mutual beneficial and cooperative theme; safety management can be structured by the help of group norms and conducted as an impressive factor with collective trainings.

Relationship Between Masculinity/Femininity and Safety Culture

In dominant masculine cultures, assertion, enthusiasm, effectiveness, competition and materialism is remarkable; differences between sexes are visible. On the other hand, in extremely feminine cultures, relations and quality of life is more important and both sexes have equal rights and responsibilities. (4)

Studies show that M/F may affect safety culture to some extent. For instance, in studies examining the effects of national culture on determining organizational safety (Tharaldsen et al. 2010) it is claimed that masculine cultures have a more calculative approach rather than feminine ones.

In Hofstede's study (1980), in masculine cultures, "respect" is mainly needed. Author explains this theoretically with the necessity of individuals' realizing their responsibilities towards themselves in masculine cultures (Demir and Okan 2009). In conforming to safety rules or adopting safety as a culture this observation may be important. Individuals may accept obeying the safety rules personally and so feeling "safe" themselves as personal responsibility. In masculine cultures, adaptation of safety culture easier can be predicted moving from here.

In masculine societies, individuals' desire to feel themselves "safe" because of their high personal responsibilities may result in their avoidance from individual safety risks as much as they can. This means they may behave more cautious and comply with rules and procedures that will provide them safety. In masculine cultures, instead of caring for others, individual success, development and getting material benefits are remarkable values that shape the behaviours. For individuals from feminine cultures, it will be a priority to give value to the human beings and relations and care for others' health and safety (Mearns and Yule 2009). Employees who come from feminine culture evaluate providing environmental health, workplace safety and colleague's safety in the same responsibility conscious, not only related to their own safety.

In order to determine an efficient safety culture in organizations, superiors should have different approaches towards individuals from M/F backgrounds.

Even though cultural influences may remain strong in a person, another strong influencer of behaviour is related to upbringing and previous experiences and learnings. These will always exert their unseen influences in our daily life, pushing us to decisions that may appear to be most appropriate as we perceive the reality with our already "biased" mind. For example, a person who has grown up in the rough and tumble of a hilly terrain area will find climbing heights or jumping off cliffs a simple task that can be achieved without fear. This person is now working inside a ship's hold at a great height. Their previous experience will prevent them from seeing "high risk" etched all over this activity. Whereas, a person who has come from a reasonable sheltered existence and not exposed to high risk activities, will find climbing up to great heights as a terrifying or dangerous activity. This dichotomy in perceiving the risk by two different persons within the same activity is to be clearly addressed to prevent complacency in taking the correct precautions for safety. When teams are at work, this variance of appreciating risks is clearly a danger and can be avoided only by good tool box meetings and trainings to understand these issues.

To sum up, to understand the term "Human element," we have to delve into the complex labyrinths of the human mind and the various influences that had molded it over the years by which an outcome (human decision) is generated. When multiple players are involved as in an organization, it can then be imagined how exasperatingly complex this process will be and how human behaviours can sometimes be totally difficult to predict even though the best trainings and processes are applied. However, a better understanding of our cultural dimensions and hidden mental processes along with its various imperfections can give rise to a better awareness of where we can go wrong in our best laid plans. This will involve pausing and becoming mindful of the way our brain nudges us to take decisions and what mysterious influences affect these inner processes.

In the next chapters, we will look at how these inner mental processes are subject to deep rooted influences that can create complex outcomes, sometimes with unpredictable consequences.

Chapter 5

Is BBS a Panacea?

People's behaviour makes sense if you think about it in terms of their goals, needs, and motives

– Thomas Mann

What Is BBS?

Behavior-Based Safety (BBS) is a program used to inform employees of their overall safety performance. It was founded on the belief that workers can be motivated to behave safely using positive reinforcement. BBS focuses on the actions and behaviors of individual employees. It puts the responsibility of safety on the shoulders of all employees, rather than just management alone.

To be successful, a BBS program must include all employees, from the CEO to the front-line workers including contractors and sub-contractors. To achieve changes in behaviour, a change in policy, procedures and/or systems will also need some change. Those changes cannot be done without buy-in and support from all involved in making those decisions.

The Basics of an Observation

Based on the operant conditioning principles developed by prominent psychologist B. F. Skinner, Behavioural Based Safety (BBS) works on the basis that behaviours that are encouraged through positive reinforcement (the application of a positive stimulus) or negative reinforcement (the removal of an aversive stimulus) are more are more likely to occur again. A positive reinforcement encourages the likelihood of a particular behaviour occurring. On the other hand, punishment (the application of an aversive stimulus) decreases the likelihood of a particular behaviour occurring again.

Thus, the main tenet of BBS is that reinforcement and punishment can be used through the provision of feedback and consequences to influence safety behaviour at work.

A combination of positive and negative reinforcement can simultaneously encourage safe- and discourage unsafe behaviours in the workplace.

When implementing a BBS program, observers (employees trained to conduct on-site safety reviews) conduct reviews of other employees by watching their behaviour. These observers record safe and unsafe behaviours, in addition to noting safe and unsafe workplace conditions. The observer then shares the findings with the worker and provides feedback. Positive feedback is encouraged. Discussing the ways in which employees can perform their tasks in a safer manner helps workers and observers to become more aware of their behaviour. BBS programs are based on a continuous feedback loop where employees and observers provide input on improving safety to each other and safety professionals utilize the data collected in conducting the observations to continually improve the BBS program.

How Should We Motivate Behavior?

Why Don't People Just Follow the Rules? (1)

Dr. Rod Gutierrez, Principal Psychologist, DuPont Sustainable Solutions gives an explanation of safety management beyond behaviour-based safety.

He states that in order to reduce lost time injuries and fatalities, various interventions were put in place. Among them were the development of processes and rules as defences against the inherent hazards and the provision of tools and safety equipment.

However, by the mid-1970s it was apparent that no major improvement could be expected from the sole deployment of these methodologies.

Enter the field of psychology, which was called upon to provide some answers on the human side of safety and provide the new thinking that was needed to push safety management towards the next millennium.

The theoretical principles underlying BBS do provide an effective and powerful solution to shaping employee behaviour and encouraging safety in the workplace. Notwithstanding the significant initial successes of BBS, it has been observed that such entirely behavioural interventions have limited effectiveness over the longer term and have not sustained a continued reduction in safety incidents; this is commonly known as the BBS plateau. A number of explanations for the limitations of BBS have been proposed.

The first limitation of a BBS approach is due to habituation, an inbuilt ability to adjust to our environment following prolonged exposure to it and which is the tendency of living organisms to cease responding to stimuli in the environment that are repetitive and iterative.

Many everyday stimuli (familiar surroundings we encounter, the pressure of the clothes on our body surface) are both repetitive and in the scheme of things unimportant. Thus, we tend to habituate to our environment and the stimuli around us as well as to the consequences applied. Similarly, over time we tend to habituate to BBS systems, safety signs and regulations found in our workplaces. Another issue of contention with the BBS approach is that it relies on the external application and internal expectation of potential consequences as the main driving mechanism for behaviour change – these consequences are delivered by an external mechanism, a supervisor, a peer a safety officer.

Thus, a purely behavioural approach is driven externally to the individual largely bypassing the complexities of personal decision making and choice selection involved in the cognitive processing. In many ways under a BBS approach, individuals are motivated to act safety by fear of repercussion and consequence rather than by a true commitment to safety as an internal value.

Another criticism of the BBS process was its focus on employee behaviour, rather than that of their managers.

Further, BBS is most effectively implemented through a very specific regime of frequency and regularity of reinforcement. These can sometimes lose vigour and consistency as managers change in the organization.

In addition, there are other critiques of BBS programs: (2)

- **BBS Programs are Difficult to Maintain.** In order to be effective, a BBS program needs full support from top tier management. It also needs to be consistently utilized and evaluated. BBS will not be effective if it's just "implemented" but lacks structure, dedication, and follow-up.

- **BBS Programs Inadvertently Place "Blame" on Employees.** While putting "blame" on employees is not the intent of a true BBS program, it's difficult to separate it out. After all, behaviour-based safety is supposed to focus on the actions and behaviours of individuals.

- **BBS Programs can Result in Inaccurate Reporting.** Because the program is structured to reward "good" behaviour, accidents and injuries can go unreported. Nobody wants to be the one who breaks the "days without injury" streak. Additionally, employees are not keen on investigations and in-depth conversations when things go wrong or when unsafe behaviour is observed.

- **BBS Programs Often Identify the Wrong "Root Cause."** When unsafe behaviours are observed, or when injuries do occur, BBS requires incident investigations. But often these investigations focus on what happened instead of the root cause, or why it happened.

- Performing observations and allowing employees to conduct those observations does not necessarily lead to changes in the way people behave at work. In most instances, it only changes the way they behave when they are being observed.

- The importance of positive reinforcement is not fully understood or implemented properly leading to lack of success in replacing unsafe work habits with safe habits.

- Not targeting management in the program for behavioural changes and not just the employees.

- Management abdication of responsibility for safety as this is primarily an employees' responsibility of observation and recording.

- Mistaken perception that behaviour-based safety is a sequence of activities, meetings, observations and data reviews, rather than a process for changing behaviour.

BBS is essentially a 'bottom up' approach as attention is directed at specific safety-related behaviours primarily performed at the lower levels of the organisation. In contrast, approaches based on the effects of social influence, are more "top down," as their focus is on understanding and changing the fundamental values and beliefs of the organisation, which are generally set or perpetuated by the leadership of an organisation.

The holistic model incorporates both top-down and bottom-up approaches, combining them to shape the cognitive patterns of individuals. BBS still has an important role to play in managing safety but incorporating cognitive elements are critical to a more sustainable and advanced system of management of employee safety. In the following chapters, we will see how the seat of all thinking, our brain, takes stock of the barrage of information coming in and how it is affected by both the external world of stimuli and the internal world of cognitive influences.

PART 2
I Think, Therefore I am

CHAPTER 6

A Peek Into the Human Brain

"The brain is a wonderful organ: it starts working the moment you get up in the morning and does not stop until you get into the office."

– Robert Frost

If all mental processes are happening in the inner realms of the human brain, then that is where our next journey should be to decipher how it influences our thoughts and how we use it for generating our decisions and guide our lives. Without doubt, the brain is the most important and complex organ in our bodies. While this book is not meant to be a textbook of biology and the author is not an expert on biology or brain sciences, a short and basic introduction is necessary to appreciate the complexities of our organ that houses the thought factory. This information is available publicly on various public sources and has been included here to highlight the importance of this highly influential tool in our thinking process and the myriad reasons why we can easily err in this, sometimes leading to catastrophic outcomes.

Regions of the Brain (1)

The brain has many different parts. The brain also has specific areas that do certain types of work. These areas are called lobes. One lobe works with your eyes when watching a movie. There is a lobe that is controlling your legs and arms when running and kicking a soccer ball. There are two lobes that are involved with reading and writing. Your memories of a favourite event are kept by the same lobe that helps you on a math test. The brain is controlling all of these things and a lot more. Use the map below to take a tour of the regions in the brain and learn what they control in your body.

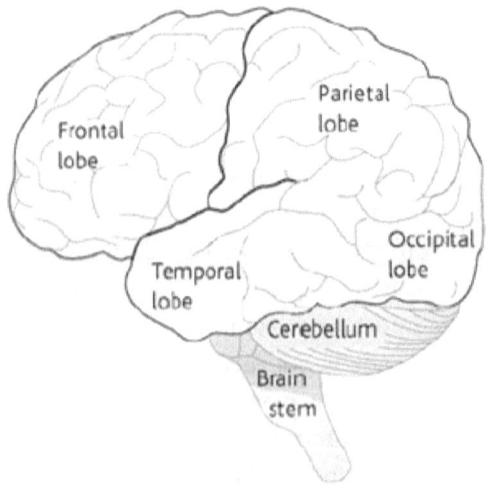

Fig. 6.1: Parts of the brain and functions

The brain is the control centre for the body. It runs your organs such as your heart and lungs. It is also busy working with other parts of your body. All of your senses – sight, smell, hearing, touch, and taste – depend on your brain. Tasting food with the sensors on your tongue is only possible if the signals from your taste buds are sent to the brain. Once in the brain, the signals are decoded. The sweet flavor of an orange is only sweet if the brain tells you it is.

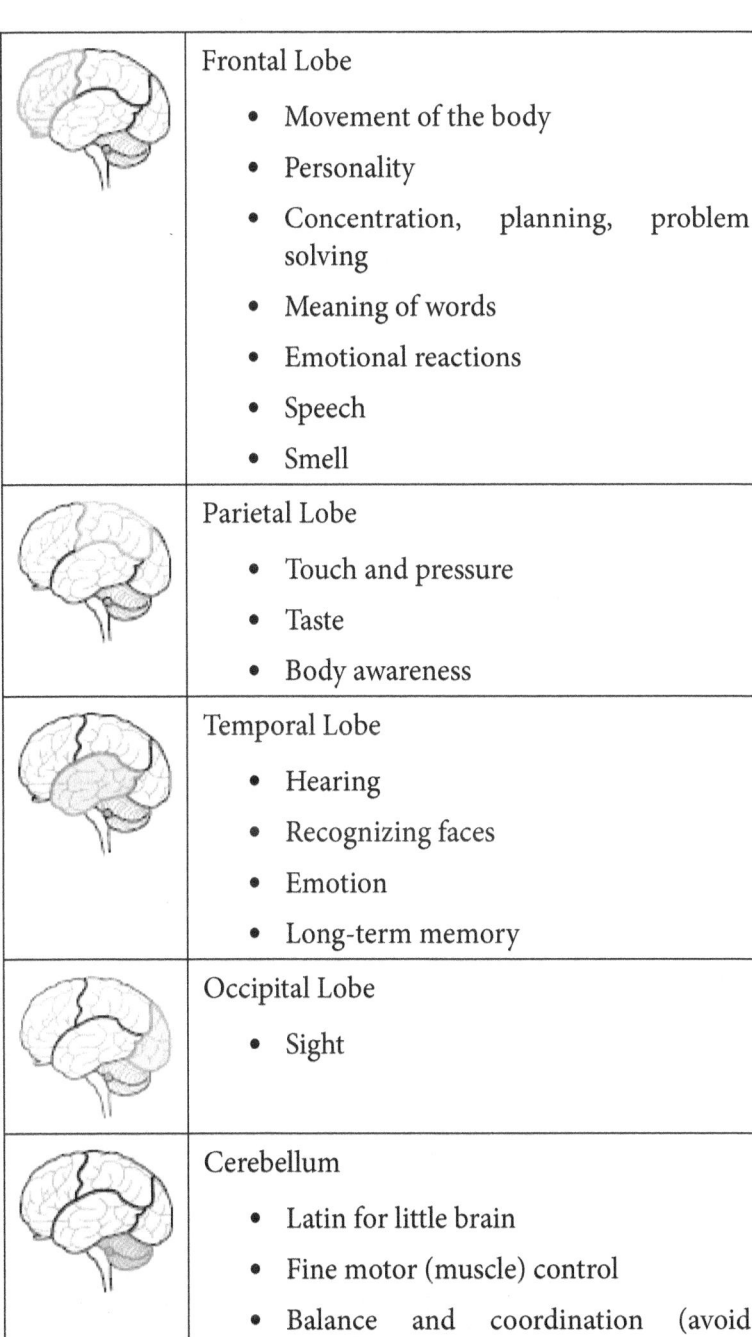

	Frontal Lobe • Movement of the body • Personality • Concentration, planning, problem solving • Meaning of words • Emotional reactions • Speech • Smell
	Parietal Lobe • Touch and pressure • Taste • Body awareness
	Temporal Lobe • Hearing • Recognizing faces • Emotion • Long-term memory
	Occipital Lobe • Sight
	Cerebellum • Latin for little brain • Fine motor (muscle) control • Balance and coordination (avoid objects and keep from falling)

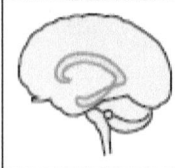	Limbic Lobe • Controls emotions like happiness, sadness, and love

BRAIN CELL

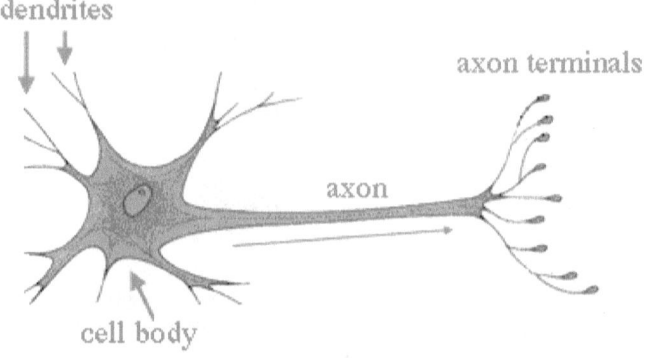

Fig. 6.2: A typical brain cell-Neuron

Fig. 6.2 shows a model of a typical type of brain cell called a neuron. A neuron is a brain cell that carries an electrical signal. Your brain contains *billions* of brain cells.

A brain cell consists of:

- A **cell body,** which stores the DNA and other things that the cell needs to do its job;

- **Dendrites,** which receive chemical signals from other cells; and

- An **axon,** which carries an electrical signal from the cell body to the axon terminals. The axon terminals contain chemicals, called "neurotransmitters," which are released in order for the cell to communicate with nearby cells.

BRAIN REGIONS

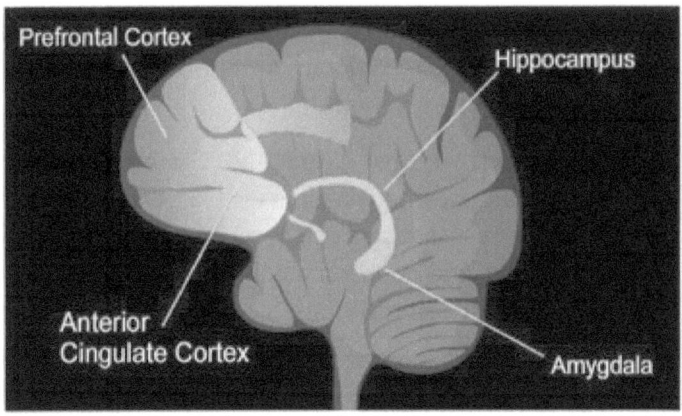

Fig. 6.3: Brain regions

Just as many neurons working together form a circuit, many circuits working together forms specialized brain systems. We have many specialized brain systems that work across specific brain regions to help us talk, help us make sense of what we see, and help us to solve a problem. Some of the regions most commonly studied in mental health research are listed below.

- **Amygdala**—The brain's "fear hub," which activates our natural "fight-or-flight" response to confront or escape from a dangerous situation. The amygdala also appears to be involved in learning to fear an event, such as touching a hot stove, and learning not to fear, such as overcoming a fear of spiders.

- **Prefrontal cortex (PFC)** —Seat of the brain's executive functions, such as judgment, decision making, and problem solving. Different parts of the PFC are involved in using short-term or "working" memory and in retrieving long-term memories. This area of the brain also helps to control the amygdala during stressful events.

- **Anterior cingulate cortex (ACC)**— the ACC has many different roles, from controlling blood pressure and heart rate to responding when we sense a mistake, helping us feel motivated and stay focused on a task, and managing proper emotional reactions.
- **Hippocampus**—Helps create and file new memories. When the hippocampus is damaged, a person can't create new memories, but can still remember past events and learned skills, and carry on a conversation, all which rely on different parts of the brain.

How Does the Brain Work?

The brain's 100 billion neurons connect the various organs and brain regions into a complex network of circuits that control specific functions within the body. These circuits serve as on/off switches for the millions of messages and processes carried out on a daily basis. The human brain is undeniably an energy system, and its very life depends on ingesting energy and then using it to fuel its activities, including complex psychological processes. The human brain consumes 20% of the body's calories even though it constitutes only 2% of the body's mass (Dunbar, 1998) **(2)**

CONCEPT OF BOTTOM BRAIN AND THE TOP BRAIN (3)

The brain is compartmentalised for our understanding, but, all the parts work in complex, intertwined ways. The largest section of the brain and closest to the surface is the Cerebrum or Cortex. This is often broken down into lobes or sections of the brain: the frontal lobe, parietal lobe, temporal lobes, and occipital lobe. The frontal lobe is at the front of the head and is responsible for planning, organisation, logical thinking, reasoning, and managing emotions. This is the part you will hear about most regarding the expression and regulation of emotions and behaviours. It is also known as the "higher brain," "rational brain," or the "top brain."

The "bottom brain" is the part of our brain that makes us act without thinking. It must do this quickly for survival purposes – if you are in a life-threatening situation, you don't have time to sit down and draw up a plan of action, you just need to act! Developmentally, this part of the brain is well developed at birth and forms more connections earlier than the top brain because it is responsible for essential tasks such as making sure our needs are met, feeling strong emotions, using instinct to keep us safe, and managing bodily functions. In some literature, it is also called as the "lizard" brain.

The top, rational brain is highly sophisticated and responsible for problem solving, rational thinking, logic, planning and decision making, organisation, and self-control. All of these things are learnt through repeated experiences. The top, rational brain is under major construction for the first few years of life. It continues to grow through adolescence, when it gets a spurt of changes until the mid-twenties when it finally matures.

The top and bottom brains are connected by a pathway on which messengers can run up and down sharing information. In our brains, these connections are less obvious but function in a similar manner. We need the bottom "emotional brain" to be able to inform the top "rational brain" with instincts and reflexes, feelings, and information about our bodily functions such as breathing, temperature, etc. However, we also need messages going from the top brain to the bottom so we can moderate and make sense of the information coming from down below.

So, what does all of this mean when we are dealing with emotions and behaviours? Basically, our kids (and some adults) are functioning primarily from the emotional, reactive, bottom brain – they are prone to sudden impulsive actions, throwing a tantrum because they got an item wrongly ordered, get frustrated with their siblings and be terrified of seeing hairy spiders. Their rational, self-controlling, top brain is still learning how to manage these situations, and the pathway between the upper and lower brains is being blocked by emotional overload so very little problem solving and effective decision making can occur. Having some understanding of the brain and how it affects our emotions and behaviours is useful. (4)

THE ROLE OF NEUROTRANSMITTERS

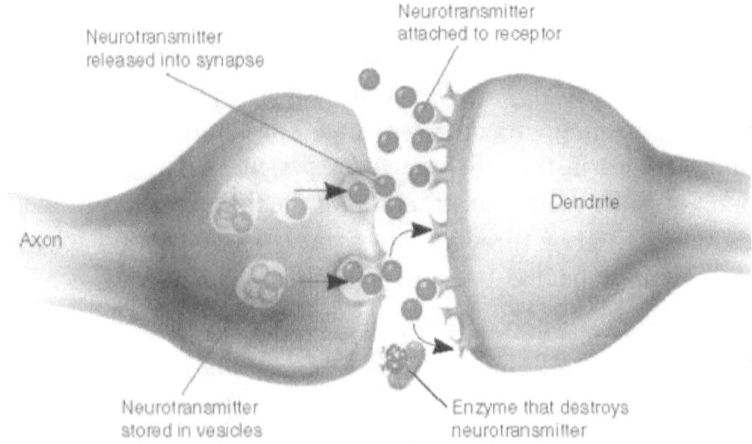

Fig. 6.4: How Neurotransmitters function

The human brain is composed of roughly 86 billion nerve cells or neurons. These brain cells communicate with each other via chemical messengers called neurotransmitters (See Fig. 6.4). For a chemical to qualify as a neurotransmitter, it must meet certain criteria. It must be found and produced inside the brain and there must be receptors specific for it alone. A simple analogy for the system of neurotransmitters and receptors is that it works somewhat like a lock and key.

Types of Neurotransmitters (5)

Scientists have found just over a hundred of these brain chemicals, and it's believed that ultimately thousands will be discovered. There is more than one way to categorize the known neurotransmitters. One common way to classify them is by their chemical *structure*. This puts them into categories like amino acids, peptides, and monoamines. Another way they are often categorized is by their *function* — whether they are **excitatory** or **inhibitory**.

What Neurotransmitters Do

Neurotransmitters regulate your mood, motivation, cravings, energy, libido, and sleep. They control your ability to focus, concentrate, learn, remember and handle stress. In many ways, they shape how you live your life and who you are. They send messages to your autonomic nervous system which controls involuntary actions like breathing, heart rate, and digestion. Abnormal neurotransmitter activity is responsible for many nervous system diseases and psychiatric disorders.

How Neurotransmitters Work

Neurons do not touch each other; there is a microscopic space between them. Thousands of neurotransmitter molecules are packaged into small sacks on the ends of neurons and get released into this space. Neurotransmitter molecules then have the potential to bind with receptors on the adjacent neuron. This is how these cells communicate with each other.

BALANCING NEUROTRANSMITTERS TO TAKE CONTROL OF YOUR LIFE (6)

An imbalance of neurotransmitters can cause problems with mood, memory, addictions, energy, and sleep. Lifestyle factors certainly play a big role in this. Chronic stress, poor diet, environmental toxins, alcohol, nicotine, caffeine, and recreational drugs are major culprits.

Underlying health conditions such as hormone imbalances, chronic inflammation, thyroid diseases, and blood sugar disorders can also cause neurotransmitter imbalances. You can also be genetically predisposed to certain neurotransmitter imbalances.

And lastly, there are entire classes of prescription drugs that work by altering neurotransmitter levels.

For example, drugs that are dopaminergic work by increasing dopamine activity, while those that are anticholinergic block the synthesis of acetylcholine.

What Neurotransmitter "Imbalance" or "Deficiency" Really Means

We hear about "neurotransmitter imbalances," "low neurotransmitter levels," or "neurotransmitter deficiencies." It generally means one or more of the following is taking place:

- Too little of the neurotransmitter is being made or formation is being inhibited.
- There are too few receptors for the neurotransmitter to bind with.
- The neurotransmitter receptors aren't working very well.
- The neurotransmitter is being broken down too soon.
- The neurotransmitter is not being appropriately recirculated.

The "Big Four" Neurotransmitters and How to Balance Them

Just a handful of all the body's neurotransmitters do most of the work. While all neurotransmitters are important, the "big four" are **Serotonin, Dopamine, Acetylcholine,** and

GABA. These are responsible for most mood disorders and problems with memory and focus. But if you don't know which neurotransmitters you need to boost, you might well be taking substances that are not doing you any good and could even be making your imbalance worse.

Here's an overview of each of the four major neurotransmitters, including symptoms of imbalances and steps you can take to optimize your neurotransmitter levels.

a. Serotonin: The "Happiness Molecule"

Of all the neurotransmitters, serotonin definitely gets the most attention. Serotonin is called the "happiness molecule" because it's so essential for a positive mood. Low serotonin is considered to be the culprit in depression and social anxiety disorders. Serotonin is also heavily involved with our perception of reality, and psychedelic drugs primarily operate on the serotonin pathways.

Low serotonin levels are linked to the most common mood disorders of our time:

- depression
- anxiety
- eating disorders
- insomnia
- obsessive-compulsive disorder
- seasonal affective disorder

Symptoms of low serotonin are:

- Feelings of unhappiness, dissatisfaction, apathy, frustration, or anger
- Feelings of depression
- Loss of pleasure towards things you once enjoyed

- Difficulty staying positive or feeling joy
- Low mood on cloudy days
- Less socializing
- No longer enjoying relationships, possibly isolating oneself by result
- Difficulty falling asleep, staying asleep, and feeling well rested

Men and women manifest low serotonin level symptoms somewhat differently. Women are much more likely to experience mood disorders and carb cravings, while men are more likely to be impulsive, have ADHD, and drink alcohol in excess.

How to Increase Serotonin?

Tryptophan is the amino acid precursor, or building block, of serotonin. It's found mainly in protein-rich foods like meat, eggs, fish, and dairy. So, theoretically, eating tryptophan-rich foods should raise serotonin levels, but the relationship between serotonin, tryptophan and food is not that straightforward. Unexpectedly, both tryptophan and serotonin levels drop after eating a meal containing protein.

It turns out that protein blocks the synthesis of tryptophan into serotonin. Eating carbohydrates alone — with no protein — at some of your meals or snacks allows tryptophan to enter your brain and boost serotonin levels there.

Daily exercise, sufficient sleep, and exposure to sunshine will increase serotonin levels too.

Symptoms of too high serotonin, or serotonin syndrome are:

- Dilated pupils
- Muscle rigidity

- Rapid heart rate and high blood pressure
- Agitation
- Anxiety
- Diarrhea
- Loss of coordination
- Muscle twitching or spasms
- Vomiting

b. Dopamine: The "Motivation Molecule"

Dopamine has been termed the "motivation molecule." It provides the drive and focus you need to do what needs to be done. Dopamine is so critical to motivation that dopamine-deficient lab mice become apathetic to the point where they'll literally starve even when food is readily available! Dopamine has another important role as the brain chemical in charge of the brain's pleasure-reward system.

Dopamine is released when your needs are about to be met and delivers a feeling of satisfaction when you've accomplished your goals.

If you've lost your zest for life or find yourself engaging in self-destructive behaviours to feel good, your dopamine level may be low.

Signs of Low Dopamine Include

Dopamine deficiency can manifest as a lethargic and apathetic form of depression unlike the anxiety-ridden depression usually linked to low serotonin. Low dopamine levels are the culprit behind the major disease Parkinson's, while too-high levels of dopamine is theorized to be a factor in schizophrenia.

Dopamine is associated with motivation, and is considered our "achievement" molecule.

Symptoms of low dopamine include

- Feelings of hopelessness or dread
- Low self-esteem/self-worth
- Lack of motivation
- Difficulty starting and finishing projects
- Easily losing one's temper from minor setbacks
- Difficulty managing stress
- Anger, aggressiveness, and irritability when stressed
- Tendency towards isolation
- Apathy towards friends and family
- Weight gain

The Best and Worst Ways to Increase Dopamine

Many people self-medicate with addictive substances like caffeine, alcohol, sugar, nicotine, and recreational drugs to increase dopamine. Others get their dopamine hit from behavioural excesses of all kinds — too much shopping, sex, gambling, video games, and thrill-seeking behaviours. Fortunately, addictions and risky behaviours are not the only way to increase dopamine!

You can increase dopamine naturally with the right foods, supplements, and lifestyle activities.

The amino acid tyrosine is a major building block of dopamine and must be present for dopamine synthesis. Tyrosine can be found in most animal food products.

Other than animal food products, foods that increase dopamine include:

- avocado
- green leafy vegetables
- apples
- beets
- chocolate
- oatmeal
- nuts
- seeds

Two of the most popular beverages on the planet, coffee and green tea, increase dopamine. *Healthy lifestyle activities like physical exercise and meditation increase dopamine.*

And since dopamine is released when you accomplish a goal, taking on new challenges helps raise dopamine levels. So, break down your long-range plans into short-term goals. Then, every time you tick an item off your "to do" list, you'll get a little dopamine boost.

c. Acetylcholine: The "Molecule of Memory and Learning"

Acetyl-choline is the neurotransmitter most involved with our pre-fontal cortex, known as the "human" part of our brains that truly separates us from animals. Acetyl-choline is involved with working memory, learning, and navigation.

Symptoms of low acetyl-choline are:

- Poor memory, or loss of visual, photographic, and/or verbal memory
- Loss of or poor creativity

- Poor word recall and loss of comprehension
- Difficulty with mental math
- Difficulty recognizing people and placing faces
- Slow mental responsiveness
- Poor spatial orientation, or clumsiness and tendency to bump into things
- Difficulty navigating or using directions when driving

Acetylcholine levels drop by as much as 90% in Alzheimer's patients.

Acetylcholine activity is the target of Alzheimer's drugs, which attempt to slow the progression of cognitive decline by blocking the breakdown of this brain chemical.

How to Increase Acetylcholine

If you are low in acetylcholine, you may find yourself craving fatty foods. If so, pay attention! Your brain is trying to tell you something. *The best way to increase acetylcholine is to quit your low-fat diet and start eating healthy fats.*

According to Datis Kharrazian, PhD, DHSc, author of *Why Isn't My Brain Working?*, the brain literally starts to *digest itself* for the raw materials needed to create acetylcholine when you don't provide it with enough dietary fat. The precursor to acetylcholine is **choline**, a nutrient found mainly in high-fat dairy products, fish, meat, and poultry. The best sources of choline *by far* are egg yolks and whole eggs.

The drugs like antibiotics, antihistamines, and antidepressants which are pretty commonly used destroy acetylcholine. Too high acetylcholine primarily operates by inhibiting other neurotransmitters. The symptoms of too high acetylcholine may be similar to the symptoms of too low serotonin, as they have a close balancing relationship.

d. GABA: "Nature's Valium"

GABA (gamma-aminobutyric acid) is a neurotransmitter that's been dubbed "nature's Valium" for its role in relaxation. This brain chemical normally inhibits brain activity on an as-needed basis, but when you're low in GABA, your mental functions get stuck in the "on" position.

Symptoms of low GABA are:

- Rapid or uneven heartbeat
- Heart palpitations
- Muscle tension
- Difficulty breathing or shortness of breath
- Sweaty palms
- Cold hands and feet
- Excessive worry
- Easily scared
- Out-of-body feelings
- Headaches
- Obsessive compulsive traits
- Unexplained feelings of stress, panic, and anxiety
- Feelings of dread or doom
- Tendency to expect the worst from people and scenarios
- Unexplained feelings of overwhelm
- Racing, restless thoughts
- Difficulty turning off thoughts when trying to relax
- Scattered attention, difficulty focusing on one task
- Worrying about scenarios that are unlikely to occur

- Feeling uneasy, on-edge
- Becoming easily tired or fatigued

How to Increase GABA

You may be drawn to unhealthy ways to increase GABA such as reaching for high carbohydrate foods, alcohol, or drugs to relax.

But there are healthy foods and supplements that will do the trick.

Good Food Sources of GABA Include:

- barley
- beans
- brown rice
- chestnuts
- corn
- kale
- potatoes
- spinach
- sprouted grains
- sweet potatoes
- yams

Fermented foods like unpasteurized yogurt, kefir, sauerkraut, kimchi, and miso also raise GABA levels as well as Probiotic supplements that contain Lactobacillus rhamnosus markedly improve GABA levels.

All kinds of exercise can increase GABA, but yoga in particular stands out.

What Contributes to Neurotransmitter Imbalance?

- Prolonged periods of stress can seriously deplete our levels. Our fast paced, fast food society greatly contributes to these imbalances.
- Poor diet. Neurotransmitters are made in the body from proteins and certain vitamins and minerals called co-factors. If your nutrition is poor and you don't take in enough protein, complex carbohydrates, healthy fats, vitamins, or minerals to build the neurotransmitters, an imbalance develops.
- Genetic factors, faulty metabolism, and digestive issues can impair absorption and breakdown of our food, which in turn reduces our ability to manufacture neurotransmitters.
- Toxic substances like heavy metals, pesticides, and some prescription drugs can cause permanent damage to the nerve cells that make serotonin and other neurotransmitters
- Certain drugs and substances like caffeine, alcohol, nicotine, antidepressants, and some cholesterol lowering medications deplete serotonin and other neurotransmitter levels.
- Hormone changes cause low levels of serotonin and neurotransmitter imbalances.

OTHER NEUROTRANSMITTERS

- **Norepinephrine. (Nor-adrenalin)**

    Deficiencies in norepinephrine occur in patients with Alzheimer's disease, Parkinson's disease, and Korsakoff's syndrome, a cognitive disorder associated with chronic alcoholism. These conditions all lead to memory loss and a decline in cognitive functioning. Thus, researchers believe that norepinephrine may play a role in both learning and memory. Norepinephrine is also secreted by the sympathetic nervous system throughout the body to regulate heart rate and blood pressure. Acute stress increases release of norepinephrine from sympathetic nerves and the adrenal medulla, the innermost part of the adrenal gland.

- **Glutamate**—the most common neurotransmitter, glutamate has many roles throughout the brain and nervous system. Glutamate is an excitatory transmitter: when it is released it increases the chance that the neuron will fire. This enhances the electrical flow among brain cells required for normal function and plays an important role during early brain development. It may also assist in learning and memory. Problems in making or using glutamate have been linked to many mental disorders, including autism, obsessive compulsive disorder (OCD), schizophrenia, and depression.

- **Endocannabinoids:** "The Bliss Molecule," Endocannabinoids are self-produced cannabis that work on the CB-1 and CB-2 receptors of the cannabinoid system. Anandamide (from the Sanskrit "Ananda" meaning Bliss) is the most well-known endocannabinoid. Interestingly, at least 85 different cannabinoids have been isolated from the Cannabis plant. The assumption is that each of these acts like a key that slips into a different lock of the cannabinoid system and alters perceptions and states of consciousness in various ways. It is likely that we self-produce just as many variations of endocannabinoids, but it will take neuroscientists decades to isolate them.

- **Oxytocin:** "The Bonding Molecule," Oxytocin is a hormone directly linked to human bonding and increasing trust and loyalty. In some studies, high levels of oxytocin have been correlated with romantic attachment. Some studies show if a couple is separated for a long period of time, the lack of physical contact reduces oxytocin and drives the feeling of longing to bond with that person again. But there is some debate as to whether oxytocin has the same effect on men as it does on women. In men, vasopressin (a close cousin to oxytocin) may actually be the "bonding molecule." But again, the bottom line is that skin-to-skin contact, affection, love making and intimacy are key to feeling happy.

- **Endorphin: "The Pain-Killing Molecule"** – The name Endorphin translates into "self-produced morphine." Endorphins resemble opiates in their chemical structure and have analgesic properties. Endorphins are produced by the pituitary gland and the hypothalamus during strenuous physical exertion, sexual intercourse and orgasm.

- **Adrenaline: "The Energy Molecule"** – Adrenaline, technically known as epinephrine, plays a large role in the fight or flight mechanism. The release of epinephrine is exhilarating and creates a surge in energy. Adrenaline causes an increase in heart rate, blood pressure, and works by causing less important blood vessels to constrict and increasing blood flow to larger muscles. An 'adrenaline rush' comes in times of distress or facing fearful situations. It can be triggered on demand by doing things that terrify you or being thrust into a situation that feels dangerous. You can also create an adrenaline rush by taking short rapid breaths and contracting muscles. This jolt can be healthy in small doses, especially when you need a pick me up. A surge of adrenaline makes you feel very alive. It can be an antidote for boredom, malaise and stagnation. Taking risks, and doing scary things that force you out of your comfort zone is key to maximizing your human potential. However, people often act recklessly to get an adrenaline rush.

Maintaining a Balanced Brain Chemistry

Brain chemistry affects the entire human being, including:

- Emotional states
- Mood, behavior, social attitude
- Physical
- Sleep, heart function, metabolism
- Mental
- Focus, learning ability

Neurotransmitters Can Act in an Excitatory or Inhibitory Manner

Neurotransmitters control the on/off switches of the nervous system. Some neurotransmitters are more likely to facilitate the transmission of certain messages and are considered "excitatory" = ON.

Likewise, some neurotransmitters are more likely to impede the transmission of certain messages and are considered "inhibitory" = OFF.

Excitatory

Excitatory neurotransmitters increase the likelihood that a neuron's signals are sent. Activation of excitatory neurotransmitters chemically change an electrical message. Excitatory neurotransmitters are responsible for providing energy, motivation, focus, they speed up the system.

Inhibitory

Inhibitory neurotransmitters decrease the likelihood that a neuron's signals are sent. This opposes or balances the effects of excitatory receptor activation. Inhibitory neurotransmitters are responsible for calming the mind and body, by filtering messages, slowing down the system, and inducing sleep.

EFFECTS OF NEUROTRANSMITTERS		
LEVEL OF NEUROTRANSMITTER	EXCITATORY	INHIBITORY
HIGH	AGITATION	WAKE UP TIRED
	OVER THINKING	NO ENERGY IN AFTERNOONS
	ANXIETY	NO MOTIVATION IN LIFE
LOW	FATIGUE	ANXIOUSNESS
	LACK OF FOCUS	ERRATIC THOUGHTS
	LACK OF EXCITEMENT	EXCITED MIND
		TROUBLE SLEEPING

Fig. 6.5: Effects of Neurotransmitters

In above figure 6.5, the effects of high and low levels of the excitatory and inhibitory neurotransmitters are shown. A fine balance is what should be ideally aim at. A balanced nervous system is necessary to maintain optimal health. When the critical balance between the excitatory and inhibitory systems is lost, it creates a situation that increases the likelihood of a neurotransmitter-related condition developing. BALANCE = optimal health

Symptoms of Neurotransmitter Imbalance

- Anxiousness
- Appetite control
- Attention issues
- Developmental delays

- Behavioral problems
- Low mood
- Fatigue
- Low libido
- Headaches
- Mood disorders
- Sleep disorders

Without enough neurotransmitters in the system, the system as a whole does not function properly. This creates a situation ripe for the onset of disease. Neurotransmitter related conditions can manifest due to an imbalance between the excitatory, and inhibitory systems.

How Does Imbalance Happen?

A healthy nervous system is characterized by meeting two basic criteria. First, it must have sufficient levels of neurotransmitters. Secondly, the excitatory and inhibitory systems must work together.

An imbalanced system is characterized by low levels of neurotransmitter supplies, with an excess of certain neurotransmitters. This imbalances and low stores of neurotransmitters are linked to:

- Anxiety
- Depression
- ADD/ADHD
- Insomnia
- Fatigue
- Weight Issues
- Obsessive Compulsive Disorder
- Migraines

Causes of Neurotransmitter Imbalance

- Hormone imbalance
- Toxins
- Poor dietary habits
- Genetics
- Stress

Due to stress, and eating poorly, excitatory neurotransmitter stores are depleted, resulting in a lack of focus, and motivation. When stress is reduced, along with healthy eating, neurotransmitter stores are replenished and balanced, eliminating depression.

This is why medications MAY or MAY NOT work. May standard prescription medication may affect one or few of the 15 different neurotransmitters. So, it may or may not work if the cause is a different neurotransmitter. Only testing can find out the cause. It is important to carry out testing to find out the true nature of the neurotransmitter imbalance and then administer any medications.

SELF-CONTROL AND GLUCOSE IN BRAIN

A famous experiment was conducted known as the "Marshmallow test" by the psychologist Walter Mischel at Stanford University in the 1970s. In this, children were called into a room one by one and told that they could eat one marshmallow now, or wait for some more time and then have two marshmallows to eat later.

Some couldn't control themselves and ate then and there and some waited for a longer period after which they were rewarded with two marshmallows.

These kids were followed up after the experiment. Years later, what sort of people did the children become?

Those with no self-control grew up with greater likelihood of drug use, criminal behaviour, poor academic performance. Those with good self-control had greater successes in life.

The brain requires tons of energy—and new experiments show how low glucose levels and self-control issues are connected. Our willpower is tested many a time in our waking hours to act or continue to act in a precise way. Or to remain in a disciplined inactivity at other times. Usually, it is in activity such as resisting a temptation, than in some observable action that the will power is tested. It turns out, however, that the power part of "willpower" is no mere figure of speech. The brain requires tons of energy—at rest, it consumes about 20% of your circulating glucose, despite constituting only about 2% of your body weight. (7)

As you carry out a particular behavior, the rate of glucose consumption jumps in the pertinent brain region. If you listen to a symphony, your auditory cortex elevates the metabolic rate. If you learn something new, it's the hippocampus that fires up. Tap dancing sparks the motor cortex. And when you're displaying willpower, thinking, "Hold on, don't give up easily – you'll regret it…" it's your frontal cortex that kicks into gear.

Work by numerous scientists, most notably Roy Baumeister of Florida State University, shows how literal the power behind willpower is.

Let's say a navigator on a routine bridge watch in an area with high traffic density encounters a vessel coming head on; he/she studies the relative movements of the vessels and calculates the right quantum of action to stay clear. As soon as the action is executed, the navigator sees another vessel on the horizon also leading to a close quarters situation. He/she works out the new safe path to navigate away from this new threat and pushes his will power to do a good job. As soon as the vessel is on the next safe path, they encounter a fishing boat looming up ahead and the navigator forces his brain to work out the next safe path to avoid a collision. These various actions sap the will power of the navigator as his/her "cognitive load" is very high and has consumed up a lot of the glucose fuelling his mental processes. And not to forget that he/she had to also carry out other incidental tasks related to the watch-keeping duties. This increase of the "cognitive load" on someone's frontal cortex, may lead to the person exhibiting less self-control on subsequent tasks—just like muscle that's been exercising hard, then balks at having to move you one step more.

Moreover, during a tough self-control task, circulating glucose levels plummet, consumed by hardworking frontal neurons. And, remarkably, self-control improves if subjects sip sugary drinks during the task (with control subjects consuming sugar-free drinks).

So, if self-control requires energy (and therefore glucose), can researchers trace the brain's struggle to control aggressive impulses by looking for low glucose levels? Brad Bushman of Ohio State University and colleagues explored this issue in a paper published recently in the Proceedings of the National Academy of Sciences.

Other experiments to track this relationship between glucose levels and aggressive impulses found that this translates into person-to-person behavior as well? It may be concluded that having a carbohydrate snack or a sweet biscuit before willpower busting situations (navigation in a dense traffic area for example) may provide for a calmer and relaxed duty execution.

More broadly, the effect of low blood glucose on willpower and judgment fits into a larger story about how stress has the same bad effects (and disrupts frontal cortical function) when it comes to violent behaviour as well. The largest lesson is that who we are and what we do must always be considered in the context of the biology occurring inside us.

The present work suggests that self-control relies on glucose as a limited energy source. Laboratory tests of self-control (i.e., the Stroop task, thought suppression, emotion regulation, attention control) and of social behaviours (i.e., helping behaviour, coping with thoughts of death, stifling prejudice during an interracial interaction) showed that (a) acts of self-control reduced blood glucose levels, (b) low levels of blood glucose after an initial self-control task predicted poor performance on a subsequent self-control task, and (c) initial acts of self-control impaired performance on subsequent self-control tasks, but consuming a glucose drink eliminated these impairments. Self-control requires a certain amount of glucose to operate unimpaired. A single act of self-control causes glucose to drop below optimal levels, thereby impairing subsequent attempts at self-control.

The scientific literature does not endorse the theory that increasing your glucose intake will all of a sudden make you more disciplined and self-controlled. We must be careful not to make this inference from the available evidence. The scientific literature only seems to endorse the idea that having enough blood glucose in the body will allow us to return to our baseline level of self-control.

This information about some of the brain functions mentioned in this chapter is important to assimilate as we are now able to drill down further when we see certain types of deviant behaviour and need not lump it as "human error." Instead, specific causes can be found if properly investigated. In most cases, corrective actions may be initiated instead of prescribing further training as a solution. And when enough case studies are investigated, we may be able to come up with an understanding of the new deeper causes of human error related incidents and probably put in preventive actions, which could also be better screenings or tests to identify and correct neurochemical imbalances.

CHAPTER 7
Stress, Anxiety and Fatigue

"My anxiety doesn't come from thinking about the future but from wanting to control it."

– Hugh Prather, **"Notes to Myself"**

Stress is everywhere and is such an oft repeated word that it appears to be more benign than what should be. An entire wellness industry has come up to solve this great "problem" of humanity. It is now commonplace knowledge that stress is a primordial response of humans to threats and a survival tool also known as the "fight or "flight" syndrome. Without it, we as a human race would have been sitting ducks to a lot of evolutionary threats and wouldn't have gone on to become the dominant species today. So, let us first demystify it by stating that it is a healthy response when it is activated as per need. However, what the modern human suffers is the deleterious effects of excessive stress on his or her health when the stress response is not managed correctly.

In historical times, the hunter ran or fought when the stress response hit his body, thereby dissipating the flow of hormones and neurotransmitters that flooded his system. Thus, any damages to his or her inner health was avoided caused by the rush of excess hormones. Today, when a ship navigator undergoes the stress response when dealing with high density traffic or any other stressful situations on board, he or she cannot physically dissipate the flow of hormonal juices that flood the system as there is never any strenuous physical activity following the stimuli causing the stress. Instead, they just use their fingers to type in a keyboard or twirl knobs on panels of the navigation equipment or other similar actions. So, the hormonal flood caused by the recent high stress situation is not dissipated due to absence of some strong physical activity. This flood of chemicals surging through the body system doesn't find a legitimate outlet and creates various adverse effects on our innards. It becomes a silent destroyer of health and leaves the chronic sufferer in a highly-strung mental state. However, stress also has other effects that potentially affects the way any person sees the world around them.

Effect of Stress on the Brain (1)

- **Stress Shrinks Your Brain**

    A 2012 Yale University study showed that chronic stress can actually reduce brain volume. In other words, if you are stressed out all the time, your brain just might shrink. Among its many effects, lower brain volume can lead to impaired cognition and hampered emotional function.

- **Stress Kills Brain Cells**

    Stress can also kill brain cells, particularly in the areas associated with memory and learning, according to the Harold and Margaret Milliken Hatch Laboratory of Neuroendocrinology, Rockefeller University. When your brain perceives stress, your body releases adrenalin into the bloodstream, giving your brain the burst of energy, it needs to fight, flee, or freeze.

    After about one minute, the adrenalin leaves your brain and it returns to normal function. If the threat persists or is severe, your brain calls in the big guns. According to Stanford University brain researcher, Robert

M. Sapolsky, these are a class of steroidal hormones, called glucocorticoids. You may have heard of cortisol, which is one example of a glucocorticoid. These hormones remain in your brain, and continue to impact its functioning, far longer than adrenalin.

Both types of hormones head straight for the hippocampus. Your body requires a state of homeostasis (hormonal balance) to function properly. The balance between sympathetic and parasympathetic hormones is very delicate. Chronic stress can keep these hormones unbalanced. High levels of either type of hormones can kill cells in the hippocampus, hampering memory and learning.

- **Stress Hinders Cognition, Expression, and Memory**

According to the University of Maryland Medical Centre, stressful events activate the hypothalamic-pituitary-adrenal system, which release catecholamines – neurotransmitters such as dopamine, norepinephrine, and epinephrine. These chemicals activate the amygdala and suppress concentration, short-term memory, rational thought, and inhibition.

This suppression of normal brain responses allows you to engage fully in fight or flight without your brain intruding. Unfortunately, suppression of such responses over the long-term can also harm your memory and impair cognitive function.

- **Stress Depletes Brain Chemicals**

If a threat is especially severe or recurs frequently, such as protracted combat or living with abuse, the chemicals that carry messages from one nerve cell to another become depleted, and the brain becomes sluggish and inefficient. According to the American Institute of Stress, this can lead to a variety of mental health effects, including:

- Depression
- Sleep disturbances
- Racing thoughts and difficulty concentrating
- Difficulty learning
- Absent-mindedness
- Difficulty making decisions
- Obsessive or compulsive behaviours
- Increased hostility, worry and guilt

- **Stress Depletes Neurotransmitters**

  In handling daily stress, the brain uses feel good transmitters called endorphins (opioids). When large amounts are needed to handle stress, the RATIO of many of the other transmitters, one to another, becomes upset creating a chemical imbalance. We begin to "feel" stress more acutely - a sense of urgency and anxiety creates more stress. Harmful chemicals are released in our bodies that do damage, causing more stress. We call this vicious cycle the "stress cycle." Emotional fatigue can result and be experienced and felt as depression.

- **Stress Surpasses Logical Thinking**

  A parent who is a non-swimmer jumps into a river when they see their child drowning. Those heroic rescues are done by people without thinking of their own safety have gone into an "auto-pilot" mode. They experienced extreme stress at seeing the situation direly needing their help, and that information entered their brain with such force and urgency that it bypassed the prefrontal cortex and other higher brain centres. It went directly into the limbic system (or the lizard brain), which is the seat of emotion and reaction. This is also the part of the brain that controls heartbeats, breathing and other autonomous bodily functions.

When incoming perceptions bypass the rational part of the brain, a person may act without the regulating effect of wisdom, memory or judgment. Logic is simply unavailable in the moment, and reactions are based on raw emotion.

- **Stress Shuts Down Functions and Sharpens Others**

  Have you ever encountered extreme stress and suddenly engaged in very instinctual behaviours? This happens because adrenaline causes the brain to start and stop certain functions in order to effectively deal with the perceived threat triggered by stress. According to the Franklin Institute, adrenaline will start and stop the following brain functions while it is present:

  - It causes the brain to signal the release of glucose and fatty acids into the bloodstream in order to provide instant energy and strength.
  - It sharpens your senses in order to perceive further threats.
  - It makes you less sensitive to pain so you can continue your fight or flight.
  - It shuts down functions unnecessary for either fight or flight, including growth, reproduction, or immunity.
  - It instructs the brain to reduce blood flow to the skin.

- **Stress Compromises the Blood-Brain Barrier**

  The blood-brain barrier protects your brain and central nervous system from viruses, drugs and chemicals that enter your bloodstream. Multiple studies, including one printed in the British Medical Journal suggest that extreme stress compromises the blood-brain barrier, making it more permeable. This makes the brain more susceptible to infection and dangerous changes in brain chemistry.

If certain chemicals can easily permeate the blood-brain barrier, you may be more susceptible to nerve damage or even brain damage.

You can, however, take steps to reduce chronic stress levels and protect your brain. Engaging in stress-reduction techniques like meditation and regular exercise can help, as can maintaining a supportive social network. If chronic stress is a part of your life, try to find ways to minimize your stress load.

Anxiety and the Brain: An Introduction (2)

It should come as little surprise that your brain is the source of your anxiety. Not only does anxiety manifest itself in thoughts – it also affects your brain chemistry in a way that can alter future thoughts and affect the way your entire body operates.

Anxiety is a natural, normal response to potential threats, which puts your body into a heightened state of awareness.

When felt appropriately, anxiety is beneficial and can keep you out of harm's way…the anxiety you may feel while hiking near a steep drop-off, for instance, will cause you to be more careful and purposeful in your movements.

While many believe anxiety and stress to be the same, persistent anxiety actually evokes quite a different experience in your brain.

- **Anxiety in Your Brain: What Happens When Anxiety Attacks?**

  Anxiety does evoke the same "fight or flight" response that stress does, which means, like stress, anxiety will trigger a flood of stress hormones like cortisol designed to enhance your speed, reflexes, heart rate, and circulation. However, stress can occur with feelings of anger, sadness, or even happiness and excitement.

Anxiety, on the other hand, virtually always involves a sense of fear, dread, or apprehension. And while stress may occur due to an external source (like an argument with your spouse), anxiety tends to be a more internal response.

Further, brief anxiety may coincide with a stressful event (such as speaking in public), but an anxiety disorder will persist for months even when there's no clear reason to be anxious. While the exact causes for anxiety disorders are unknown, your brain is actively involved.

The National Institute of Mental Health explains:

"Several parts of the brain are key actors in the production of fear and anxiety...scientists have discovered that the amygdala and the hippocampus play significant roles in most anxiety disorders.

The amygdala is an almond-shaped structure deep in the brain that is believed to be a communications hub between the parts of the brain that process incoming sensory signals and the parts that interpret these signals. It can alert the rest of the brain that a threat is present and trigger a fear or anxiety response.

The emotional memories stored in the central part of the amygdala may play a role in anxiety disorders involving very distinct fears, such as fears of dogs, spiders, or flying. The hippocampus is the part of the brain that encodes threatening events into memories."

- **Neurotransmitters and Anxiety**

Many neurotransmitters have been linked to anxiety, including:

- Serotonin
- GABA
- Norepinephrine

Even dopamine may play a role in anxiety, or at least have a calming effect on those already living with anxiety symptoms. Interestingly, too much or too little of any hormone may also affect anxiety in different ways. The problem is with balance. If your brain doesn't have enough serotonin, for example, it may cause you to experience anxiety symptoms.

When it comes to neurotransmitter production, the truth is that cause and effect are rarely known. It's often impossible to distinguish between poor neurotransmitter balance as a result of life experience, or poor neurotransmitter balance as a result of genetics. Both can occur in anyone living with anxiety, and in some cases a combination of both may be responsible for anxiety symptoms.

- **Anxiety and Brain Activation**

There are two different parts to an anxiety disorder, and someone with anxiety may suffer from one or both. The first part is mental – verbal worries, nervous thoughts, etc. The second part of anxiety is physical. For example, a racing heartbeat, panic attacks, light headedness, and other physical symptoms.

It's possible to experience physical symptoms with less worry, and it's possible to worry often without many physical symptoms. Researchers also found that both excited different parts of the brain. Those with worried thoughts showed more left-brain activity when nervous. Those with physical symptoms experienced more right brain activity.

Another study looked at the way that those with a spider phobia reacted to the belief that they were going to encounter a spider. They found that those with the phobia had their dorsal anterior cingulate cortex (ACC), insula, and thalamus become more active than those without a phobia.

Yet another study at the University of Wisconsin – Madison found that those with generalized anxiety disorder appeared to have a weaker connection between the white matter area of the brain and the pre-frontal and anterior cortex. This was compared to those without generalized anxiety disorder and the results appeared to be significant.

These are just some of the ways that anxiety can activate the brain.

- **Hormones and Anxiety**

Hormone balances may affect anxiety as well. Many different hormones have an effect on brain chemistry and neurotransmitter production and balance, so if these hormones appear to be out of balance, anxiety may be the result.

Some examples of hormones affecting the brain include:

- **Adrenaline/Epinephrine** – Adrenaline is one of the most common causes of anxiety symptoms. Your body releases it when your fight or flight system is active, and it causes the increase in heart rate, muscle tension, and more. In some cases, long term stress and anxiety may damage your ability to control adrenaline, leading to further anxiety symptoms.

- **Thyroid Hormone** – Thyroid hormone appears to regulate the amount of serotonin, norepinephrine, and Gamma-aminobutyric acid (GABA) produced and distributed to the brain, so problems with your thyroid may also increase your risk for developing anxiety.

Several hormones may cause anxiety, and a change in brain chemistry may increase the production of hormones that lead to further anxiety symptoms.

- **Panic Attacks and the Brain**

    Panic attacks are a particularly distressing form of anxiety, and these may be due to the health of the brain too. Researchers have found that those with panic attacks often have an overactive amygdala. While it's not clear what creates this over activity, the fact that that area of the brain appears to contribute to panic attacks indicates that some aspect of the brain is in control of the panic attack experience.

- **Other Links Between Anxiety and the Brain**

    Another interesting relationship between anxiety and the brain is that long term anxiety may damage the brain in a way that could cause further anxiety. Researchers have found that when you leave your anxiety disorder untreated, the dorsomedial prefrontal cortex, anterior cingulate, hippocampus, dorsolateral prefrontal cortex, and orbitofrontal cortex all appear to decrease in size. The longer the anxiety goes untreated, the smaller and weaker they appear to be.

    What's interesting is that not only do these changes affect anxiety symptoms – they also create anxious thoughts. Those with anxiety may feel their thoughts are completely natural, whereas, the brain contributes to that type of negative thinking.

Chronic Fatigue Syndrome (3)

While acute fatigue has been studied very closely and many shipping companies have come up with Fatigue Management Plans to tackle it, Chronic Fatigue Syndrome (CFS) has not been addressed in the same manner. It is the most common name used to specify a disorder or group of disorders generally defined by persistent fatigue unrelated to exertion, not substantially relieved by rest and accompanied by other specific symptoms for a minimum of six months.

The most common primary characteristic of Chronic Fatigue Syndrome is excessive and persistent physical and mental fatigue.

Chronic Fatigue Syndrome is a Limbic System condition brought on by neurological trauma which may involve viral, environmental and/or psychological factors. The way in which the brain/mind/body expresses this trauma is unique to each person. Trauma can also initiate an inflammatory response in the central nervous system.

The condition of Chronic Fatigue indicates that the brain is stuck in a distorted self-protective mechanism centred on energy conservation. This cross-wired neuronal circuitry directly affects the physiology of the body and manifests in a range of neurological, immunological and endocrine system abnormalities.

In response to a chronic trauma cycle, the body's abilities to rest, digest or regenerate are affected interrupting the normal growth cycle and detoxification process catapulting the brain and body into a cycle of chronic illness.

Symptoms of CFS include:

- post-exertional malaise;
- unrefreshing sleep;
- widespread muscle and joint pain,
- cognitive difficulties,
- brain fog,

- chronic, often severe, mental and physical exhaustion and other characteristic symptoms in a previously healthy and active person.

CFS patients may report additional symptoms including muscle weakness, hypersensitivity, orthostatic intolerance, digestive disturbances, poor immune response, and cardiac and respiratory problems, mild fever, sore throat, tender neck with swollen lymph nodes. CFS is often accompanied by depression and isolating behaviours.

Once Limbic system function is corrected, the destructive cycle is discontinued and energy levels and cognitive function begin to return to normal.

Muscle and joint pain and weakness subside, and the central nervous system inflammatory cycle is interrupted.

The hyper vigilance in the systems of the body relaxes which allows the natural rest and healing processes to resume.

- **A Vital Test in Chronic Fatigue Syndrome**

    The central problem of chronic fatigue syndrome is mitochondrial failure resulting in poor production of ATP. ATP is the currency of energy in the body and if the production of this is impaired then all cellular processes will go slow. The perfect test is to measure the rate at which ATP is recycled in cells and this test has now been developed by John McLaren Howard. He calls it "ATP profiles." It is a test of mitochondrial function.

    Not only does this test measure the rate at which ATP is made, it also looks at where the problem lies. Production of ATP is highly dependent on magnesium status and the first part of the test studies this aspect.

    The second aspect of the test measures the efficiency with which ATP is made from ADP. If this is abnormal then this could be as a result of magnesium deficiency, of low levels of Co-enzyme Q10, low levels of vitamin B3 (nicotinamide adenine dinucleotide NAD) or of acetyl L-carnitine.

The third possibility is that the protein which transports ATP and ADP across mitochondrial membrane is impaired and this is also measured.

The joy of the ATP profiles test is that we now have an objective test of chronic fatigue syndrome which clearly shows this illness has a physical basis. This test clearly shows that cognitive behaviour therapy, graded exercise and anti-depressants are irrelevant in addressing the root cause of this illness.

CHAPTER 8

Distortions of Our Reality

"Our mental limitations prevent us from accepting our mental limitations."

– Robert A. Burton

Seafarers nowadays carry out most type of routine and non-routine duties by completing the necessary checklists, permits and related documentation to ensure that all processes and safety considerations are followed. In addition, a risk assessment including MOC (Management of Change) wherever needed are completed. This should to a large extent ensure that the work in question will be safely and effectively executed. Also, the very fact that these precautions are taken will give the workers a sense of confidence and seriousness while carrying out the job.

And prior to the actual work, the supervisor calls a Toolbox meeting to discuss the work in detail and the various hazards and controls in place. With all these measures in place, it is expected that there should be no safety issues during this activity. But it is still surprising that accidents or serious near-misses occur despite these precautions. So where are we going wrong? The issues could be in any of the following:

- Poor compliance with checklists. Tick culture strongly prevalent and the setting in of a state of complacency.

- Permits to work not supervised sufficiently or not thorough.

- Risk assessment process not creative enough to identify various hazards. Generic hazard list being used faithfully to "complete" the RA.

- Toolbox meeting is just a formality and a rushed affair. Team not fully engaged in perceiving the risk & reality properly. Supervisor's idea of reality and risk prevailing with no further discussions.

- Checklists, Permits and Risk Assessment process are leading to a sense of complacency in team members.

- Reasons beyond the control of vessel staff.

Other reasons could be lack of knowledge, poor training, poor teamwork and communication, fatigue effects, ill health, distractions etc. However, a point that is generally ignored and usually spoken about only in upper echelons of management is the differing manner of each individual perceiving the world around him or her because of the human brain construct. With trillions of neurons in our brain, and the dazzling variety of thought generation possible for the same stimuli, no two humans are likely to think alike in similar situations. Also, no two humans will generate the same neuronal outcome for the same set of instruction. Hence, this fact is to be kept in mind closely when looking at team dynamics and expected team outcomes. This chapter aims to highlight that these effects on the brain are for real and "my reality" need not be "your reality." You and I are different due to various genetic and environmental influences which have been ingrained over many years of evolution, upbringing and past experiences.

On any given day, how often do you think that you are in total control of all your faculties? If you are well rested, the answer is that you feel confident that you will quickly understand the situation at hand and can take reasonably correct actions to complete the operations competently and safely. However, are you fully in control of all your faculties? And are you perceiving the reality around you as you "intended" it to be? It may come as a surprise that you may be reacting to a world of make believe in which the culprit is of your own making...your own brain playing tricks on your so-called sensible mind. Let us look at how we perceive our world of reality and how distortions can deceive this reality!

Our Brain Plays Tricks On Us

Our brain is a wonderful and amazing thing, but it certainly isn't perfect. Sometimes it forgets essential details like your spouse's birthday or sometimes the exact time of a doctor's appointment. Other times it fails to notice stuff right in front of you or essential things in the world around you, leading to mistakes or poor judgments.

One might shrug it off as simple errors or blame situational causes. You were too busy, too tired, or perhaps too distracted. The fact is, however, that your brain has several limitations and patterns that can trip you up in several different ways. It is always processing tons of information coming through its senses and billions of neurons in your brain are working together to generate a conscious experience. And not just any conscious experience, your experience of the world around you and of yourself within it. How does this happen? According to neuroscientist Anil Seth, we're all hallucinating all the time; when we agree about our hallucinations, we call it "reality."(1)

The following are just a few of the psychological tendencies that might lead you astray. (2)

1. Our Brain Likes to Take Shortcuts

One of the biggest shortcomings of our brain is that sometimes it's behaves as if it is plain lazy. This is because the brain tries to always economize on its energy consumption and evolutionary needs dictated it to take shortcuts. When trying to solve a problem or make a decision, our mind often falls back on rules of thumb, mental shortcuts, or solutions that have worked well in the past. In many cases, this is a useful and effective approach.

Using such mental shortcuts allows one to make decisions quickly without having to laboriously think through every possible solution. But sometimes these mental shortcuts, known as heuristics, can trip us up and cause us to make mistakes.

For example, you might find yourself terrified of flying on a plane because you can immediately think of several tragic, high profile plane crashes. In reality, traveling by air is actually much safer than traveling by car, but because your brain is using a mental shortcut known as the availability heuristic, you are fooled into believing that flying is much more dangerous than it is in reality.

Stereotyping is a type of heuristic that all people use to form opinions or make judgments about things they have never seen or experienced.

2. Hidden Biases Influence Your Thinking (3)

We are also susceptible to a number of cognitive biases that prevent us from thinking clearly and making accurate decisions. These common mental mistakes are patterns of thinking that result in errors, distortions, and downright inaccurate conclusions.

Cognitive biases are patterns of thinking that don't rely on logic. Because they allow your mind to get around gaps in knowledge and memory, you need them to bypass things like extreme pressure, time constraints, emotions, and overwhelming information. However, because cognitive biases are based on generalizations and assumptions, they can't always be correct. And if you don't check your reasoning, they can lead to judgements and decisions that negatively impact your business. You can't eliminate them all, but you can become more aware of how they function and ways to counteract them.

The best protection is awareness. Forewarned is forearmed. If you cannot eradicate the distortions ingrained in the way your mind works, you can at least build tests and disciplines into your decision-making process that can uncover errors in thinking before they become errors in judgment.

List of Some Common Types of Cognitive Bias

1. **Overconfidence Bias**

   Overconfidence results from someone's false sense of their skill, talent, or self-belief. It can be a dangerous bias and is very prolific in maritime industry. The most common manifestations of overconfidence include the illusion of control, timing optimism, and the desirability effect. (The desirability effect is the belief that something will happen because you want it to.)

2. **Self-Serving Bias**

   The self-serving bias refers to our tendency to take personal credit for success while blaming outside sources for our failures. Essentially, we tend to believe that our successes are due to internal traits and talents, while our failures are caused by variables outside of our control.

3. *Herd Mentality*

   Herd mentality is when people blindly copy and follow what other role models or famous persons are doing. When they do this, they are being influenced by emotion, rather than by independent analysis. There are four main types: self-deception, heuristic simplification, emotion, and social bias.

4. *Framing Cognitive Bias*

   Framing is when someone decides because of the way information is presented to them, rather than based just on the facts. In other words, if someone sees the same facts presented in a different way, they are likely to come to a different conclusion about the information. If the same familiar hazards are always used for a RA, then the resultant risk identification will give different controls instead of the needed ones.

5. **Narrative Fallacy**

   The narrative fallacy occurs because we naturally like stories and find them easier to make sense of and relate to. It means we can be prone to choose fewer desirable outcomes due to the fact they have a better story behind them. This cognitive bias is like the framing bias.

6. **Anchoring Bias**

   Anchoring is the idea that we use pre-existing data as a reference point for all subsequent data, which can skew our decision-making processes. If you see a phone that costs $1,000 and then another phone that costs $500, you could be influenced to think the second phone is very cheap. Whereas, if you saw a $200 phone first and the $500 one second, you might think it's very expensive.

7. **Confirmation Bias**

   Confirmation bias is the idea that people seek out information and data that confirms their pre-existing ideas. They tend to ignore contrary information. Confirmation biases impact how people gather information, but they also influence how people interpret and recall information. For example, people who support or oppose a particular issue will not only seek information that supports their beliefs, they will also interpret news stories in a way that upholds their existing ideas and remember things in a way that also reinforces these attitudes.

8. **Hindsight Bias**

   Hindsight bias is the theory that when people predict a correct outcome, they wrongly believe that they "knew it all along."

9. **Representativeness Heuristic**

   Representativeness heuristic is a cognitive bias that happens when people falsely believe that if two objects are similar, then they are also correlated with each other. That is not always the case.

10. **Halo/Horns Effect:**

    The halo effect is a type of cognitive bias in which our overall impression of a person influences how we feel and think about his or her character. Essentially, your overall impression of a person ("He is nice!") impacts your evaluations of that person's specific traits ("He is also intelligent!"). The horns effect is the opposite.

11. **Recency Bias**

    This is another variation on the halo/horns effect and refers to the tendency for an assessor to be influenced by a person's most recent performance – no matter how they might have performed earlier.

12. **Self-Serving Bias**

    If you ace an exam, it's because you studied hard. If you failed, on the other hand, it's because the teacher didn't explain the subject properly, or the classroom was too warm, or your roommate kept you up all night before the exam.

13. **Leniency Bias**

    This refers to an assessor's tendency to rate people higher than they merit. Such ratings may be motivated by the assessor's aversion to confronting poor performance, by friendship or sympathy, or because they fear the impact of poor ratings on team motivation

14. **Simplification Bias**

    In the interests of efficiency over thoroughness an assessor may attribute success or failure to what people do at the time, rather than seeing their behaviour as the intersection of a great many decisions at many organisational levels, most of them made much earlier.

15. **Attentional Bias**

    When you are trying to make an important decision, do you always consider all of the possibilities? While one might like to think that we take all the alternatives into consideration, the reality is that we often overlook some options and possible outcomes. In some cases, our attention becomes focused on just a few of the options while we ignore the rest. This tendency represents a type of cognitive bias known as an attentional bias.

    The attentional bias can also have an impact on memories. Since people can become overly focused on a single stimulus, they might neglect to notice other aspects of the situation. When recollecting the event later on, memories may be distorted, inaccurate, or incomplete due to this bias.

16. **The Sunk-Cost Trap**

    Another deep-seated bias is our tendency to make choices in ways that justify past decisions, even when the latter no longer seem valid. We know rationally that sunk costs are irrelevant to present decisions, but they nevertheless prey on our minds and lead to inappropriate choices.

    This frequently occurs when we're unwilling, consciously or not, to admit a mistake. Acknowledging a poor business decision is a very public matter, inviting criticism from colleagues and bosses. It's psychologically safer to justify past decisions, make allowances and continue in the present course, even when we know the outcome is risky. You'll need to make a conscious effort to set aside any sunk costs — psychological or financial—that muddy your thinking.

17. **The Status-Quo Trap**

    We all carry biases that influence the choices we make. For example, each of us is predisposed to perpetuating the status quo; it's an inherent part of our thinking. Deep within our psyches, we are self-protective and risk-aversive.

18. **Change Blindness: (4)**

    There is simply so much going on in the world around us at any given moment that our brains cannot attend to every detail. As a result, we can sometimes completely miss major changes that happen right in front of our eyes. Do you think that you would notice if the person you were talking to suddenly switched into someone else mid-conversation? Researchers have found that when conversational partners were swapped during a brief interruption, the majority of people didn't even notice the change. Change blindness refers to this failure to detect differences in visual scenes.

    So why are we so prone to missing important shifts in the world around us? Researchers suggest that several different factors probably play a role. First, we have to deal with the limited resources that are available to us. If we are busy concentrating on one thing, we simply have to tune out huge amounts of other information that our brains cannot deal with.

3. Our Memory Isn't as Great as We think

While we often believe that our memory works like a video camera, carefully preserving events exactly as they occurred, the reality is that our memory is much more fragile, inaccurate, and susceptible to influence than we would like to believe.

Your memory might be good, but it is worth remembering that it is not perfect and certainly not always dependable.

We also tend to forget enormous amounts of information, from trivial details that we encounter each and every day to important information that we need. Our memory of events is usually sharper if we have associated events with things or other events to make the recall easier. Also, the remembering context of the situation is very important when we try and recall them.

If the above mentioned issues with our cognitive process is not enough to convince the reader that we live in a world of make believe reality due to our various thinking defects, then the next sections will look at how the visual stimuli that that make us believe in our external world (I see, so I believe!) can also play tricks on our helpless brain.

Distortions and Visual Perception (5)

Many of the most familiar optical illusions are *distortions*. They take advantage of the brain's assumptions to skew the way you perceive contours, lengths, colours, and shading.

For example, the long diagonal lines in the following picture (which run from the top-left to bottom-right) are perfectly parallel. However, the pattern of cross marks in the line fools your brain into thinking they lean toward one another.

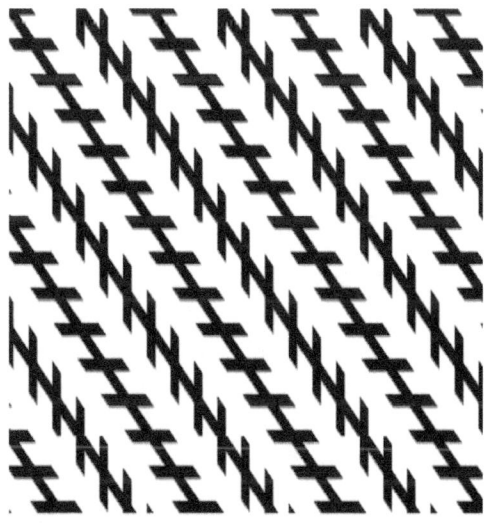

Fig. 8.1

The Zöllner Illusion was created by Johann Karl Friedrich Zöllner (1834–1882), a German astrophysicist with a keen interest in optical illusions. Here your brain is confused by angles that aren't quite what it expects. It's as if your brain expects the hatch marks to cross each line at a right angle. You can almost feel your brain mentally twisting the lines to make them fit its expectation.

Fig. 8.2

Similarly, in the above figure, the brain distorts the lines to adjust to the poorly aligned blue and yellow patterns creating the impression that the lines are not parallel to each other, which in fact it is.

The following image shows a more ambitious pattern that easily blinds the brain. The image shows a series of concentric circles, but the brain is locked into a different interpretation, and insists on seeing a spiral. (Trace your finger around one of the circles if you don't believe it's con centric.)

Fig. 8.3

The remarkable part of both these illusions isn't that your brain is fooled—after all, its mistaken logic is reasonable and (more importantly) it's blindingly fast. The amazing part is that even if you carefully measure the angle of the slanted lines or trace out the circles, thereby proving the illusion, you still can't convince your brain that it's made a mistake. In fact, no amount of pleading can convince your brain to alter its wonky interpretation. Your brain may take a lot of rules into account when it decides how to view a scene, but it has no interest in your slow-thinking deductive logic.

Fig. 8.4

In one look, the brain will only see 4 legs for the elephant. However, on a closer look, the brain will have trouble calculating the number of legs.

Faulty Comparisons

Along with distortions of shape, your brain can also mislead you when sizing up the length, size, and colour of an object. And when the brain's assumptions fail, the effects tell us quite a bit about the brain's book of visual rules, tricks, and shortcuts.

For example, the following illusion shows two curved shapes. The bottom shape appears to be larger, but it's actually identical to the top shape.

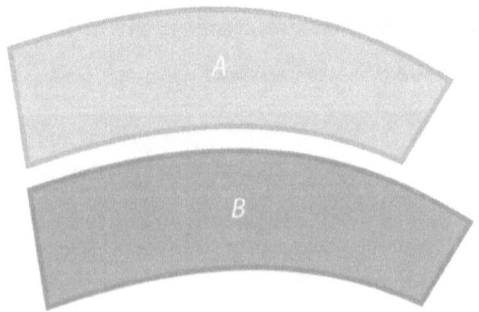

Fig. 8.5

This illusion works because in its haste, your brain makes a few simplifications. It notices the way the left edges of both shapes line up, and takes that into account, discounting the fact that the lined-up edge is gently slanted. When your brain then turns its gaze to the right side, it correctly notices that the bottom shape kicks out a bit further. Thus, the brain concludes that the bottom shape is bigger, missing the fact that the left edge of the bottom shape actually sits a bit further to the right than the edge of the top shape. (If the shapes were truly lined up, the top-left corner of each shape would be positioned on the same vertical line.)

A similar faulty rule is on display in the orange circles of the next optical illusion.

Fig. 8.6

Here, your brain makes two correct observations: the orange circle on the left is small compared to its blue neighbours, and its counterpart on the right is large compared to its neighbours. However, once the brain settles on this intriguing fact, it becomes blind to the fact that both orange circles are the same size. Instead, the proportionally larger one (on the right) seems larger than the one on the left.

The distortion and faulty comparison illusions are among some of the most profitable for business. These illusions lie behind the tapered packages of shampoo and ice cream that are constantly being reworked to look just as large while holding ever less.

Here are a few of many products with packaging that can betray you:

- Condiment bottles with ridiculously long necks. The brain is better at judging size (the area a shape takes on your eyeball) than volume (the space a container actually has to hold ketchup).

- Bottles of maple syrup that bulge out pleasantly in the middle (where you're most likely to look) but narrow dramatically at the base.

- Sticks of antiperspirant that tower impressively high, while being whittled down to a thickness of a few microns.

- Packages with multiples of anything. Often these packages use carefully designed windows to show you some of the items inside, and use artful contours to imply there are more items inside. When you open the package, you find less than you expect. Your brain's expectation is based on what will fit in the package, but the package designer is more interested in maximizing profit than efficiency.

- Gift baskets supported with vast quantities of unseen tissue paper. Again, your brain sizes up the overall shape and size of a product when judging whether it's worth a second or third look.

When shopping, don't rely on your vision to make a final purchasing decision. Fortunately, most products are required to have key facts stamped on their packages (like weight). Although studying this information won't lead your brain to see the package any differently.

A similar effect is at work in the legendary same-colour illusion. Here, two squares that are filled with *exactly* the same shade of grey (A and B) appear to be dramatically different. Once again, it's almost impossible to accept this illusion unless you cover up almost everything else in the picture except the two squares in question.

Fig. 8.7

The remarkable part of this illusion is that the brain picks up on a range of clues to make an emphatic conclusion—everything from the 3-D shape of the cylinder that casts the shadow to the pattern of the checkerboard, which darkens significantly but imperceptibly around square B. (The latter part is the most significant factor in the illusion. The brain is deeply attached to the idea of a regular checkerboard pattern, and prefers to see that over anything else.)

Fig. 8.8: Kanizsa's Triangle (8)

A floating white triangle, which does not exist, is seen. The brain has a need to see familiar simple objects and has a tendency to create a "whole" image from individual elements. Gestalt means "form" or "shape" in German. However, another explanation of the Kanizsa's Triangle is based in evolutionary psychology and the fact that in order to survive it was important to see form and edges. The use of perceptual organization to create meaning out of stimuli is the principle behind other well-known illusions including impossible objects. Our brain makes sense of shapes and symbols putting them together like a jigsaw puzzle, formulating that which isn't there to that which is believable.

Fig. 8.9(9)

To make sense of the world it is necessary to organize incoming sensations into information which is meaningful. Gestalt psychologists believe one way this is done is by perceiving individual sensory stimuli as a meaningful whole.[16] Gestalt organization can be used to explain many illusions including the rabbit – duck illusion where the image as a whole switches back and forth from being a duck then being a rabbit and why in the figure – ground illusion the figure and ground are reversible.

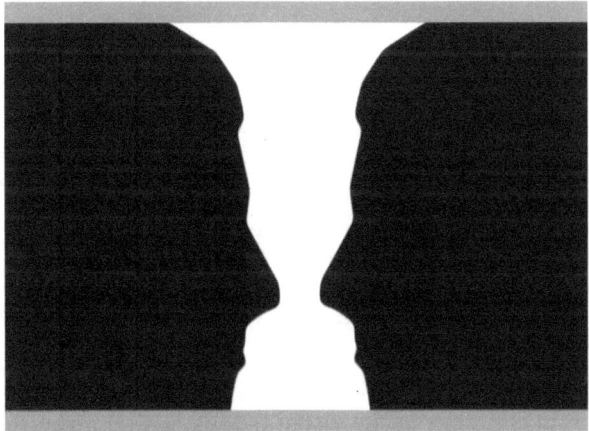

Fig. 8.10: Reversible figures and vase, or the figure-ground illusion(10)

Figure – Ground Organization is a type of perceptual grouping which is a vital necessity for recognizing objects through vision. In Gestalt psychology it is known as identifying a *figure* from the back*ground*.

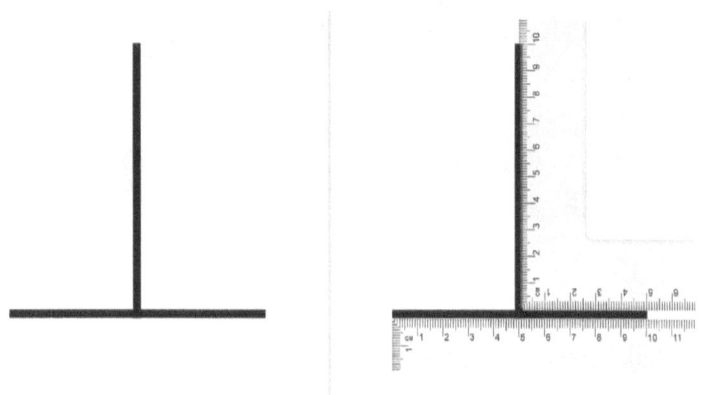

Fig. 8.11

In Fig. 8.11, the vertical – horizontal illusion where the vertical line is thought to be longer than the horizontal.

Fig. 8.12 (11)

In the Ponzo illusion the converging parallel lines tell the brain that the image higher in the visual field is farther away, therefore, the brain perceives the image to be larger, although the two images hitting the retina are the same size.

Fig. 8.13(12)

Simultaneous Contrast Illusion. The background is a color gradient and progresses from dark grey to light grey. The horizontal bar appears to progress from light grey to dark grey, but is in fact just one colour.

Fig. 8.14 (13)

"Shepard's tables" deconstructed. The two tabletops appear to be different, but they are the same size and shape.

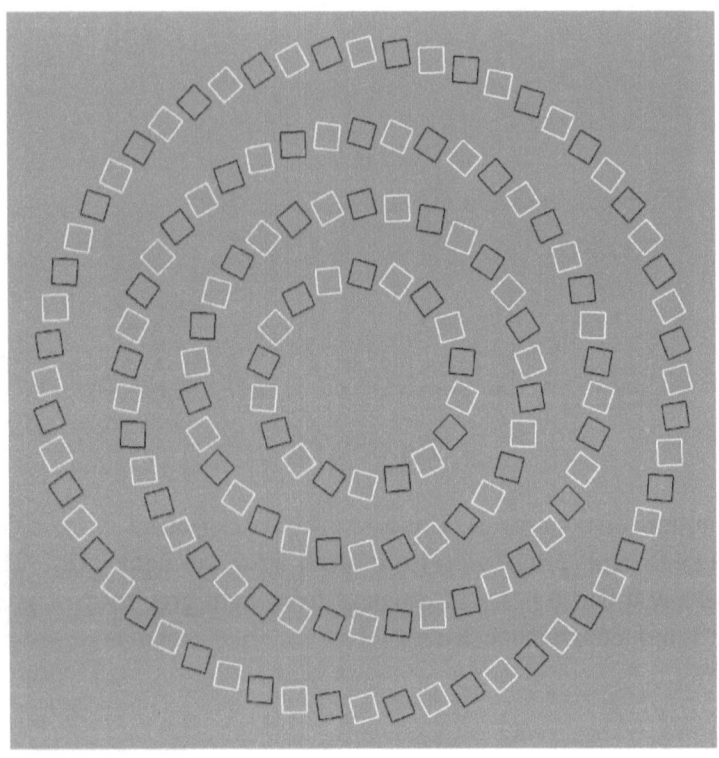

Fig. 8.15 (14)

Pinna's illusory intertwining effect and Pinna illusion. (The picture shows squares spiralling in, although they are arranged in concentric circles.)

Fig. 8.16 (15)

Scintillating grid illusion: Dark dots seem to appear and disappear rapidly at random intersections, hence the label "scintillating."

The 3-D World

So far, you've seen how the brain has built-in assumptions that help it interpret shapes, sizes, and colours (and sometimes lead to quirky mistakes). The brain also has a bag of tricks that it uses to convert the 2-D image that's projected on your eye to a realistic understanding of the 3-D world in front of you.

Consider the classic example of two lines, shown below. Even though a ruler will tell you that the lines are the same length, the brain stubbornly insists that the top one is shorter.

Fig. 8.17

One explanation for this illusion is that the brain is biased towards picking out the cues of 3-D objects. Lines that angle inward are typically seen in objects that are nearby (like the table in the picture below). Lines that angle outward are more common in distant objects (like the back corners of the room). Here's an example that illustrates by comparing two lines that have the same length, but are placed in two different spots in a 3-D scene.

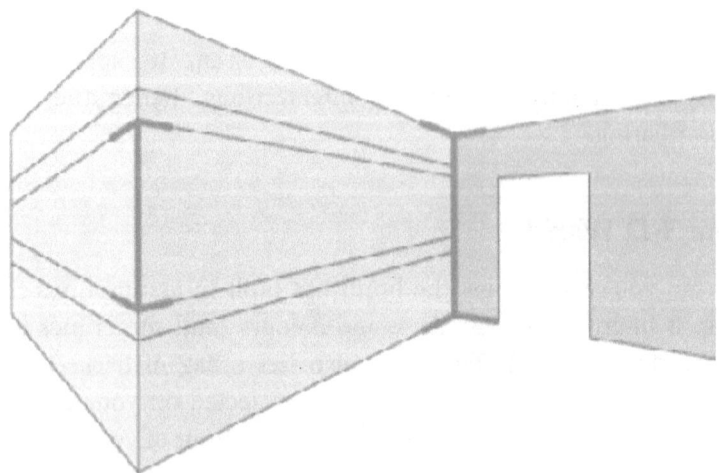

Fig. 8.18

In the two-lines illusion, your brain is well aware of the fact that both lines are really and truthfully the same length. However, your brain also believes that the bottom-most line is farther away. If two objects look the same in your eye, but one is farther away, there's only one possible conclusion—namely, the object that's farther away is bigger. Thus, the brain "corrects" the length of the second line to take the imagined distance into account.

At first, it seems odd that the brain is so willing to skew the size of things based on their perceived distance. However, on second thought it makes a lot of sense. If the brain didn't perform this automatic adjustment, your father would appear to shrink to midget size as soon as he began walking away from you.

The brain has several other tricks for translating the 2-D picture in your eye into a 3-D model. It assumes that objects close to the horizon are farther away, and it compares unknown objects against nearby known objects to infer distance.

Another 3-D cue is shading. When the brain takes in a scene, it expects to find a sun-like light source radiating from above, and it uses patterns of shading to infer contours and shapes. Humans co-opt these automatic assumptions with artful applications of makeup. To an unbiased observer (say, a computer or an alien being from another planet), makeup would seem like little more than face paint. But for the easily influenced human brain, makeup is processed like shadows, and suggests a more sharply defined face.

Lastly, the brain uses one physical detail to see in three dimensions: the slightly different vantage point that's provided by each of your two eyes. You probably already knew this, but it's a less important factor than you probably thought. The separation of your eyes helps your brain accurately judge depth for very close objects, but it's useless for far off ones. As you can readily test, if you cover one eye and wander around the house you might have trouble doing some precision tasks (like tying a knot or chopping tomatoes), but you won't have any difficulty interpreting the shapes around you as 3-D objects.

In the below image (Fig. 8.14), the text at first appears gibberish. But as we allow the brain to take in the information, the words automatically pop out with the brain fitting the right word from its memory to give meaning to each word by also solving the simple number-word puzzle. This happens at quite a fast pace that clearly shows how in the real world, the brain distorts what we see and make us see according to the context and the need to obtain meaning as per its earlier stored memories.

GOOD EXAMPLE OF A BRAIN STUDY. IF YOU
CAN READ THIS YOU HAVE A STRONG MIND.

7H15 M3554G3
53RV35 7O PR0V3
H0W 0UR M1ND5 C4N
D0 4M4Z1NG 7H1NG5!
1MPR3551V3 7H1NG5!
1N 7H3 B3G1NN1NG
17 WA5 H4RD BU7
N0W, 0N 7H15 LIN3
Y0UR M1ND 1S
R34D1NG 17
4U70M471C4LLY
W17H 0U7 3V3N
7H1NK1NG 4B0U7 17,
B3 PROUD! 0NLY
C3R741N P30PL3 C4N
R3AD 7H15.
PL3453 5H4R3 1F
U C4N R34D 7H15.

WITITUDES.COM

Styles of Distorted Thinking (6)(7)

In addition to the logical fallacies that can misrepresent or misuse evidence, here is a list and short description of several common forms of distorted thinking.

Filtering (selectivity): This is a failure to consider all the evidence in a balanced and objective assessment. We go where our attention is, and our attention is inherently limited. Selectivity is a failure to consider a neutral, or balanced, point of view. It can have two basic forms. The first is considering only the negative details and magnifying them while filtering out all the positive aspects of a situation. The second is taking the positive details and magnifying them while filtering out all the negative aspects of a situation. In any case evidence that supports your bias is selected, favoured, or weighted more heavily than evidence contrary to your bias. Find the realistic balance between the optimistic and pessimistic points of view. Seek out, carefully consider, and assimilate *all* the evidence.

Overgeneralization: It is incorrect to arrive at a general conclusion based on a single incident or piece of evidence. This is a common example of the more general fallacy of basing a conclusion on unrepresentative evidence. Consider a broad range of representative evidence before drawing a conclusion. Consider systematic evidence, and dismiss anecdotal evidence.

Polarized Thinking: This is the fallacy of thinking that things are either black or white, good or bad, all or nothing. This fallacy can lead to rigid and harmful rules based on primal thinking when it is efficient to compress complex information into simplistic categories for rapid decision-making during times of stress, conflict, or threat. Polarized thinking can also lead to unhelpful forms of perfectionism. The reality often lies in the sizeable middle ground between these extreme poles. Recognize and reject the false dichotomy. Some phenomenon is intrinsically dual.

Mind Reading: You conclude, incorrectly and without considering other alternatives or testing your assumptions, that you understand how another person is thinking and what their reasons and motives are for taking a particular action. This is an example of the Fundamental Attribution Error where you incorrectly attribute an action or intent to an agent. One example of this is drawing a negative conclusion in the absence of supporting information. Focusing only on evidence that supports a negative position, while neglecting to consider alternative positive explanations is the fallacy of not considering representative evidence. It is false to conclude the "he must hate me because he didn't say 'hi' to me." There are many plausible explanations for why he neglected to say "hi."

Personalization (Egocentric Bias): This is the fallacy of incorrectly thinking that everything people say or do is a reaction to you. It is an egocentric viewpoint where you attribute personal meaning to everything that happens. Face it, you are not really that important nor influential. This point-of-view often causes the predator to view himself as the true victim; their cause is just and is not to be thwarted. It also often results in a set of self-centred rules.

Attribution Errors: It is a fallacy to believe you can correctly know a person's intent for behaving as they do. Their actions may or may not be deliberate. The person may not even be aware of what they are doing. Their actions may or may not be directed at you. Their actions may have unintended consequences or may result from an accident or chance. We judge others based on behaviour and we judge ourselves based on intent. It is difficult to determine *cause* when only *effect* can be observed. This error is so common and so misleading it has been named the Fundamental Attribution Error (FAE).

Intentional Stance: A class of attribution errors based on the belief that outcomes only result from an agent's intent, and that bad things are the result of intentional evil. One example is attributing natural disasters such as drought, floods, and hurricanes to the revenge of supernatural forces. Intent cannot be reliably inferred from behaviour.

Pattern Discernment: We may think we see a pattern that isn't there; the outcomes are simply the result of random events. Or we can think we recognize a pattern that is different from what we actually see. We may also fail to recognize a pattern that is present.

Catastrophizing: You anticipate an unreasonable disaster based on a small problem. Every scrap of bad news turns into an inevitable tragedy. It is the error of using a personal, pervasive, and permanent explanatory style despite contrary evidence. Consider a broader range of representative evidence before drawing a conclusion. Strike a realistic balance between optimistic and pessimistic views. Skip the histrionics.

Control Fallacies: It is a fallacy to mistake what you can change for what you cannot change. Do not underestimate the degree of control you have for your own actions. You are not helpless, powerless, nor perpetually a victim. Examine the alternatives you have for taking action and responsibly for your life. Also do not overestimate your responsibility for the happiness and pain of others. Be realistic in evaluating the power and influence you do and do not have over yourself and others.

Fallacy of Fairness: Your sense of justice may not be shared widely and is certainly not shared universally. The world may not be fair, or at least it may not always work according to what you feel is fair. Examine your own sense of justice and continue to reconcile it with what happens in the world. The principle of empathy is a good basis for justice. Anger is the emotion that urges us to act on our sense of justice. Choose your battles carefully to make the most constructive use of your limited time, energy, and other resources. Don't harbour resentment at every injustice you perceive, and examine your feelings of self-righteousness. Gather evidence to make an informed decision.

Fallacy of Change: It is unrealistic to believe you can change other peoples' nature, personality, deeply ingrained habits, or strongly held beliefs. Be realistic about what you can change and what you cannot. Do not depend unrealistically on others for your own well-being.

Ignorance: Choosing to ignore or dismiss relevant information, choosing a narrow worldview; refusing to inquire, examine, study, and learn; rejecting alternative viewpoints before examining or considering them; ignoring or denying evidence; choosing to stay unaware; and holding desperately to your limited beliefs are all ways to choose ignorance over wisdom and more carefully considered evidence. Blind faith, forgetfulness, and lack of introspection are also forms of ignorance. When coupled with your attachment to an idea, belief, someone, or something, ignorance can surface as pretention, deception, shamelessness, lack of rigor, inconsideration and disrespect of others, and distraction.

Being Right (denial): Dogmatically holding onto an opinion, belief, or defending an action can be a destructive result of stubborn pride. Denial is a failure to acknowledge evidence. Even if you believe you are right, decide if you would rather be right or be happy. Don't waste time pursuing the *fallacy of change* described above. Examine your sense of justice and the assumptions you are making. Gather evidence to make an informed decision, but even if you are right, it may not be a battle worth fighting. How is this working for you now?

Cognitive Dissonance: Tension between thoughts and actions inconsistent with those thoughts. A tense and uncomfortable contradiction exists unless your actions support your thoughts and beliefs. To close the gap and relieve this tension humans often revise their thoughts to support their actions. Irrevocable bad decisions are similarly defended. People who bought the wrong car, lost money in the stock market, went on a disappointing vacation, or got a bad haircut spontaneously invent clever defences for the actions they are now stuck with. What is remarkable is how strongly we believe these self-justifying stories when we make them up ourselves.

Optimism: Believing that all is good and everything will turn out fine provides the important benefits of encouraging us to persist toward our goals and overcome obstacles. However, unchecked optimism can easily detach us from the cold harsh truths of reality. Examine the evidence, think critically, allow for scepticism, consider a variety of viewpoints, come to a balanced conclusion, and act responsibly.

Just World Theory: The mistaken belief that good things happen to good people and bad things happen to bad people. This is sometimes used as an excuse to blame the victim; "he got what he deserves."

Asch Effect: People often change their opinions to agree with the majority, despite the presence of clear contrary evidence. Experiments conducted by Solomon Asch demonstrated the effects of group pressure on the modification and distortion of individual judgment. Experimental subjects often modified their judgment or estimate of an observation to conform with the majority opinion of a group.

Stereotypes: Human memory is organized into *schema* which are clusters of knowledge or a general conceptual framework that provides expectations about events, objects, people, and situations in life. This attribute of memory leads us to rely on *stereotypes*. These are simplified and standardized conceptions or images held in common by members of a group. While stereotypes are an essential feature of human memory, they can cause problems when the attributes associated with the group are incorrectly extended to an individual. For example, a common stereotype of a bird includes the ability to fly, however extending that stereotype to a penguin leads to an incorrect conclusion.

Magical Thinking: Believing that the laws of physics, economics, or the laws of cause and effect, don't apply to you. Believing in miracles or believing that wishful thinking or sheer will alone can cause the outcome you are hoping for are examples of magical thinking, as are appeals to paranormal or supernatural phenomena. Don't let optimism exceed the bounds of reality. Hope is not a strategy.

Accepting Repetition as Evidence: Sometimes a person will simply repeat their opinion when asked to provide evidence to justify an assertion or belief they have expressed. They may repeat their position emphatically, engage in various dominance displays, highlight various power symbols, show impatience, or assert their positional power as they simply repeat their opinion. A variation of this fallacy is to claim "everyone knows . . . is true" as the evidence. But repetition is not evidence, and it should not be accepted as evidence.

Assumptions, Opinions, Rumours Become Fact: It is easy for assumptions, opinion, or rumours to be accepted as fact. This can happen if these ideas or stories seem reasonable on the surface, or they support your views or interests, if they advance some hoped for outcome, or they are expressed by someone in authority or someone you trust, if the stories are fun to tell, or if others you know also share these beliefs. The incorrect assumption, opinion, and rumour that the earth is the centre of the universe went unchallenged by millions of people for perhaps thousands of years. Other rumours and unchallenged assumptions can be even more destructive.

Final Thoughts

Our brain is capable of remarkable things, from remembering a conversation we had with a friend to solving complex logical problems. But obviously, it isn't perfect. So, what can we do? There's no way to avoid all these potential problems, but being aware of some of the biases, perceptual shortcomings, and memory tricks that our brain is susceptible to can help.

Our thinking is the result of our own perception, judgment, experience, and bias. Our brain distorts reality to increase our self-esteem through self-justification. People perceive themselves readily as the origins of good effects and reluctantly as the origins of ill effects. We present a one-sided argument to ourselves.

During times of stress, overload, or threat, we often resort to a simplistic form of thinking, called *primal thinking*, that incorporates many of these fallacies. For an accurate appraisal it is important to reassess the situation using effortful, valid, thoughtful, and accurate analysis that properly allows for the complexities we face. Employ critical thinking and work to understand *what is*.

The reader will now be convinced that while perceiving the world around us, "what you see may not be what you get." Hence everybody will have to be extremely careful in judging the reality of the world around him or her knowing that one's senses may be taking him or her for a ride by distorting the reality of our everyday world. This intelligence that we hold has great value while evaluating risks around us and enable us to "see" the picture more clearly than just blind acceptance of the sensory inputs which may lead us to great dangers while estimating hazards and their effects on us. When assessing hazards in a work area, one person may be able to "see" only part of the real "reality" of the area and so, it makes sense to use a team approach to check and confirm using the other team members. This may remove some of the distortions and biases of "seeing" by individual team members and allow a safer consensus on what the "reality" actually is. Such an approach will enhance the safety of the operation and better team performance. Risk assessments should as far as possible be never an individual exercise for the common good.

CHAPTER 9

Focusing the Brain

"Wisdom is 'seeing through the illusion"

– McKee & Barber

The brain processes all kinds of information that comes in as stimuli from the external world at the conscious and sub-conscious level. Apart from the various sensory inputs, feelings and thoughts associated with the moment are also processed to bring about a unique reality of those moments for the observer. But if the brain must process any information at a conscious level, there should be an awareness maintained on the input stimuli. This focus or spotlight beamed on the point of interest and the attention paid to it is an elusive mental faculty in the mind's operations and its role in living a fulfilling life. How we deploy our attention determines what we see. In other words, "our focus is our reality"!

Selective attention is the neural capacity to beam in on just one target while ignoring a staggering sea of incoming stimuli, each one being a potential target for focus. This is what William James (founder of modern psychology) defined attention as "the sudden taking of possession by the mind, in clear & vivid form of one of what seems several simultaneously possible objects or trains of thought."

Focus demands we tune out our emotional distractions, which means that those who focus best are relatively immune to emotional turbulence, more able to stay unflappable in a crisis & have a stable response to life's changing emotional onslaughts.

Failure to drop one focus & move on to others can leave the mind lost in repeating loops of chronic anxiety.

During sharp focus, key circuitry in the pre-frontal cortex gets into synchronized state with the object of the beam of awareness, a state referred to as "phase-locking."

George Miller proposed that 7 plus or minus 2 chunks of information as the upper limit of the beam of attention since the 1950s. Nowadays, it is estimated that four chunks of information is the upper limit.

Many people think that splitting attention is multi-tasking. Cognitive science tells us that this is fiction. Instead of splitting it, we switch rapidly. Continual switching saps attention from full concentrated engagement.

Full absorption in what we do feels good, and pleasure is the emotional marker for flow. People love what they are doing.

The optimal brain state for getting work done well is marked by greater neural harmony – a rich, well timed interconnections among diverse brain areas. When brains are in this zone, we are more likely to perform at our personal best whatever our pursuit.

On the other hand, another large group are stuck in a "frazzled" state, when constant stress overloads their nervous system with floods of cortisol & adrenaline. Their attention fixates on their worries, not their jobs. This emotional exhaustion can lead to burnout.

As described in the Chapter 6 – "A peek into the Human Brain," our brain has two semi-independent, largely separate mental systems. Daniel Goleman, in his book "Focus – The hidden driver of excellence," (1) explains that the two mental systems of the brain as "Top down" and "Bottom up" brain. The Bottom up brain has massive computing power and operates constantly, quietly solving our problems, surprising us with sudden insights or solutions. Since it operates beyond the horizon of conscious awareness, we are blind to its workings.

When talking on a cellphone while driving a car (the driving part is in the back of the mind) & a sudden honk makes you realize that the light has changed to green. This is processed by the "top-down" brain which refers to the mental activity mainly within the neo-cortex, that can monitor & impose goals on the sub-cortical machinery.

The "Bottom-up" brain is faster, involuntary, always on, intuitive, impulsive, manages the mental models of our world. In contrast, the "top down" mind is slower, voluntary, effortful, the seat of self-control, able to learn new models.

Voluntary attention, willpower and intentional choice are top-down while reflexive attention, impulse and rote habits are bottom-up brain activities. Our mind's eye is constantly dancing between voluntarily directed focus and stimuli driven attention capture.

The bottom-up mind multitasks, scanning a profusion of inputs in parallel. It analyzes what's in our perceptual field before letting us know what is selects as relevant for us. Our top-down mind takes more time to deliberate on what it gets presented with, laboring over item by item and applying more thoughtful analysis. Most of what the top down mind believes it has chosen to focus on or think about and do is plans dictated by the bottom-up systems. The bottom up circuitry dates to millions of years of evolution and favours short term thinking, impulse and quick decisions. The Top-down circuitry are later additions, probably about a few hundreds of thousands of years.

The brain's more ancient bottom-up systems apparently worked well for basic survival during most of the prehistoric days. In today's world, its effect is seen in impulsive behaviours, addictions, overspending, reckless behaviours etc.

As neuroscientists know, the brain always tends to economize on energy consumption. Top-down efforts like learning a new topic requires more energy. But the more we keep learning the same topic, it changes into a rote habit and gets taken over by the bottom-up circuitry. This part of the brain that does this is called the basal ganglia, a golf ball sized mass at the brain's bottom and above the spine. As any routine is practiced more, it gets passed on to the bottom-up circuitry. The brain is constantly dividing its work between both the systems to be energy efficient. Due to this neural transfer, over time, we stop noticing the effort to carry out the activity as the bottom up brain has gone into automatic mode.

Subconscious choices in attention are made when the amygdala circuitry, the brain's sentinel for emotional meaning, also it's radar for threat, spots something it finds significant. A large hairy spider, an angry tone in voice, a slithering snake… The bottom up system reacts faster in neural time than does the top down prefrontal area.

Our brain's attention mechanisms evolved over hundreds of thousands of years to survive in a dangerous environment where threats approached our ancestors within a specific visual range & set of rates—Those of our ancestors whose amygdala was quick enough to help us escape that lion or other threats passed on their neural design to us.

We are wired to pay reflexive attention to "super-normal stimuli"; whether for safety, nutrition or sex…like a cat that can't help chasing a fake mouse on a string. In today's world, the advertising industry uses these to play on those same pre-wired inclinations to activate us at the bottom-up, too, getting our reflexive attention. Just tie sex or scandal to a product to activate these same circuits to prime us to buy for reasons we don't even notice. Such capture from below is automatic; an involuntary choice.

When these circuits spots a threat (or what we interpret as a threat), a superhighway of neuronal circuitry running upward to the pre-frontal areas sends a barrage of signals that let the lower brain drive the upper; our attention narrows, making it easier to recall anything relevant to the threat at hand; our body goes into overdrive as a flood of stress hormones prepares our limbs to fight or run. We fixate on what's so disturbing & forget the rest. So how long does our focus stays glued? That depends on the power of the left pre-frontal area to calm the aroused amygdala (there are two amygdale, one in each brain hemisphere). When we are hijacked, the amygdala circuitry captures the right side & takes over. But the left side can send signals downward that calm the hijack.

People who are highly emotionally resilient – who bounce back from upsets right away – can have as much as thirty times more activation in the left pre-frontal area than those who are less resilient.

Active engagement of attention signifies top-down activity, an antidote to going through the day with a zombie-like automaticity. This focused goal-oriented attention inhibits mindless mental habits.

Half our thoughts are of the daydream variety. Wandering of mind may have some advantages over the seriously focusing types. The inner tug to drift away from effortful focus is so strong that cognitive scientists see the wandering of mind as a default mode.

Neuroscientists have observed that while the mind wanders, our sensory systems shut down, and, conversely, while we focus on the here and now, the neural circuits for mind wandering go dim. At the neural level, mind wandering and perceptual awareness tend to inhibit each other; internal focus on our train of thought tunes out the senses, while being rapt in the beauty of a sunset quiets the mind. This tune out can be total, as when we get utterly lost in what we are doing.

Each partial tune out bears risk; one study of a thousand drivers injured in accidents found that about half said their mind was wandering just before the accident; the more intense the disruptive thoughts, the more likely it was the driver caused the accidents.

Situations that do not demand constant task-focus – particularly boring or routine ones-free the mind to wander.

Tightly focused attention gets fatigued-much like an overworked muscle – when we push to the point of cognitive exhaustion. The signs of mental fatigue, such as drop in effectiveness and a rise in distractedness and irritability, signify that the mental effort needed to sustain focus has depleted the glucose that feeds the neural energy.

The antidote to attention fatigue is the same as for the physical kind; take a rest. But what rests a mental muscle? Try switching from the effort of top down control to more passive bottom up activities, taking a relaxing break in a restful setting. The more restful surroundings are found in nature. (Not web surfing, playing video games or answering email)

This trigger bottom up attention modestly allowing circuits for top down efforts to replenish their energy, restoring attentiveness and memory, and improving cognition.

What allows the brain to work quickly and efficiently is its energy supply. If this is impaired in any way, then the brain will go slow. Initially, the symptoms would be of foggy brain; but if symptoms progress, one will end up with dementia. We all see this in our everyday life, with the effect of alcohol being the best example. Short-term exposure gives us a deliciously foggy brain – we stop caring, we stop worrying, it alleviates anxiety. However, it also removes one's drive to do things, one's ability to remember; it impairs judgement and our ability to think clearly. Medium-term exposure results in mood-swings and anxiety (only alleviated by more alcohol). Longer-term use could result in severe depression and then dementia

You are always creating your future. You bring it forth through your thoughts, actions, feelings, beliefs, values, goals and dreams. You do this regardless of the level of your conscious awareness. Your present moment awareness coupled with the future that you create is a deeper reflection of your subconscious programming.

All your future goals and dreams are not only a reflection of your subconscious thinking, they are also mediated by your Reticular Activating System (RAS) (See Fig. 9.1). The RAS is the part of your brain that serves as a filter between your conscious mind and your subconscious mind. The RAS, which is located in the core of your brain stem, takes instructions from your conscious mind, and passes them on to your subconscious mind. The *reticular activating system* (RAS) is the portal through which nearly all information enters the brain. (Smells are the exception; they go directly into your brain's emotional area.) The RAS filters the incoming information and affects what you pay attention to, how aroused you are, and what is not going to get access to all three pounds of your brain

Because of this biological function, whatever you are thinking about or focusing upon will seep down into your subconscious mind only to reappear at a future time. Have you ever decided that you wanted to buy a certain car, and shortly thereafter, you see cars everywhere like the one you wanted? That is how the RAS works.

This function of the Ras can be put to great advantage by deliberately focusing on the desired goals for any activity. This setting of your intent plays a key role in encouraging your subconscious mind to home in on your desired goal, as well as create the most optimal future. According to the Merriam-Webster Dictionary, the word intent is derived from the word intend, which means to direct the mind and proceed on course towards a goal.

When you set your intent, you are directing your Reticular Activating System to stretch towards your desired goal and future, and to also enjoy the journey getting there.

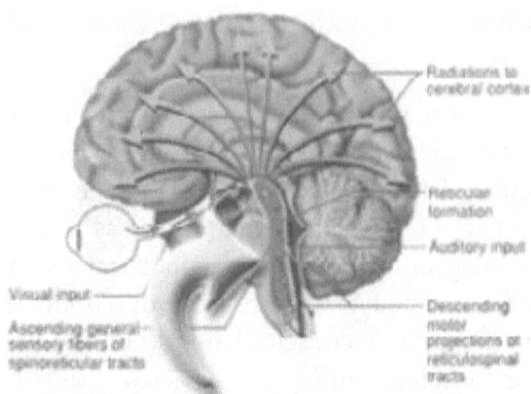

Reticular Activating System

Fig. 9.1: Reticular Activating System (RAS)

To gain an experience with setting your intent and positively programming your RAS, try saying the following three sentences to yourself:

1. "I hope to enjoy the play tonight." (Notice how you actually think about this – your internal pictures, voices, and feelings.)

2. "I want to enjoy the play tonight." (Notice how you actually think about this – your internal pictures, voices, and feelings—what is different from the first question?)

3. "I intend to enjoy my play tonight." (Notice how you actually think about this—your internal pictures, voices, and feelings—what is different from the first two questions?)

Pay attention to how each of these simple changes in your language creates a very different experience. For most people, the first question will produce some doubt. In other words, multiple images will appear in your mind representing different possibilities—one is that you may enjoy the play and the other one being that you won't.

The second sentence should produce a different representation. When you say, "I want to enjoy the play tonight," you will typically see what you want in the future, but you may not see yourself having it now. The future then may feel compelling because you see what you want. But there is still some room for doubt because it is more difficult to put yourself into the actual experience of achieving it.

The third image of intending to enjoy your play should put you into the act of fully enjoying your experience and being present to it. Intending for something to happen will generally associate you into the experience of achieving your goal and all the feelings, images and sounds that go with it.

When you set your intent, you are marrying your subconscious mind with your conscious will to make something happen. It is like you are sending your Reticular Activating System a message that you are "expecting" the event to happen, and there is absolutely no room for uncertainty.

Setting your intent is a way of preparing your subconscious mind and RAS for the kind of journey that you will have in achieving your desired goal. The setting of your goal represents the end result you want to achieve. The RAS filters information based on its degree of novelty, personal importance and whether it fits with a pattern of familiar information.

Understanding the RAS is very important to safety management as it suggests individuals may not always consciously act in an unsafe manner, particularly if their RAS is not engaged or 'switched off'. If the RAS is not processing relevant safety information an individual may not be aware of any risk in their behaviour. So, if safety is instilled as a important component of workplace culture and a firm, personal belief, the RAS may be more likely to identify it as a 'important and relevant' and the individual will be able to maintain more conscious safety. For example, during a work activity inside an enclosed space, the setting of intent will involve deliberate and conscious thinking of each of the work activity processes and imagined as being carried out safely and efficiently. This strong intent setting will assist in quickly picking out any deviations from safe conditions or acts by our RAS and will flag it as an impending threat.

Setting your intent is a powerful way of directing your conscious energy and attention towards your future goal, which in turn helps your subconscious mind and RAS stay focused on the desired outcome. Your subconscious mind and conscious mind are a system that co-exists within a larger system that we call reality. (3)

How we think, act, and behave has a direct influence on the greater system of our external reality. When we set our intent, we are influencing both our inner reality, and our outer reality in a way that sets a chain of events into motion. We are bringing forth a new chain of events that are directly related to our deeper subconscious thinking, as well as our overall intent for the desired outcome and journey that unfolds.

An individual whose RAS is 'switched off' may subconsciously still act safely because his or her safe behaviours may have been reinforced through behavior-based trainings. However, with the RAS 'switched on,' the conscious decision to act safely may be further reinforced by appropriate attention focus.

What Is Brain Fog? (4)

When workers are sent in to carry out critical work, it is imperative that the work is carried out as per good practices and a team that is mentally and physically fit for the purpose. Many a times, the mental fitness part of this requirement is overlooked as apparently, the team members seem to know what they need to do and, they have previous experience in this activity. However, what is sometimes not realized is that their mental faculties may not be at its optimum levels due to insufficiency of proper thought processes and which may not be apparent to other team members. A difficulty to concentrate, poor mental stamina, difficulty in recall, meaning something and saying something else are all signs of a condition known as "brain fog." If good energy supply to the brain is affected, the thought generation is also affected thereby leading to flawed decision making.

Good fuel and oxygen supply depend on the brain getting the right amounts of blood supply, maintaining the right blood pressure, having enough oxygen in the atmosphere and sufficient fuel (glucose). This good fuel and oxygen supply will be converted into energy that the brain can use, namely ATP (Adenosine Triphosphate) which happens in the mitochondria of the cell.

There is lots of research showing that essential fatty acids are indeed "essential" for normal brain function; so oils that would be helpful in addition to coconut oil would be omega 3 (fish), omega 6 (evening primrose) and omega 9 (olive), together with lecithin (which is phosphatidylcholine – i.e. the main component of all cell membranes).

It would be a good idea to include the following oils as part of the daily diet to bring down the effects of brain fog:

- **Lecithin** – one teaspoon (5 ml) twice daily (raw material for basic building blocks of membrane).
- **Coconut Oil** – one dessertspoonful (10 ml) twice daily (perfect fuel for brain cells).
- **Hemp Oil** which has the right proportion of omega 6 to 3 (4 to 1) (ensures membranes are of perfect consistency – not too stiff, not too elastic).

Research done by Professor Caroline Pond at Milton Keynes has shown that the immune system, like the brain, is also fat-loving. Wild animals, if they have a food glut, will first deposit fat around lymph nodes where energy is needed for immune activity.

There are other treatments for foggy brain, but it is advised that a qualified medical professional be consulted for the right kinds of treatment.

Attention Deficit (Hyperactivity) Disorder (ADD or ADHD)

This is another disorder that affects the capability to focus our brain on the activity at hand. According to experts in the field of ADD/ADHD, the disorder is the result of a neurotransmitter imbalance. Many people with ADD are highly intelligent.

Recognizing ADD

A professional diagnosis is the best way to determine ADD/ADHD in any individual. However, the following description, as given by experts in the field of ADD/ADHD, serves as a guide.

A high level of frustration causes ADD people to be impatient. Whatever is going on—they want it to go quickly and be finished. People with ADD suffer from "overload"; they have a heightened awareness of incoming environmental stimuli. Their world tends to be too bright, too loud, too abrasive and too rapidly changing for comfort.

Unable to filter out normal background "noise" they find it difficult to concentrate on a task before them. Disorientation to time and space is often a problem. They may have difficulty following a set of instructions or reading a map. ADD people tend to be disorganized. They have trouble making and carrying out plans. Many ADD people are hyperactive. As youngsters they're constantly moving, squirming, twisting and getting into everything. As adults they're restless and easily distracted. They often tend to forget appointments, to pay bills and complete tasks. Because they're always in a hurry, delays of any kind make them frantic. ADD people live under such stress, frustration is difficult to tolerate, and when they're frustrated, they're likely to become angry.

TEN BRAIN HACKS FOR IMPROVING ALERTNESS AND PERFORMANCE (5)

#1 – Vitamin D

Vitamin D not only protects neurons, but also regulates enzymes in your brain and cerebrospinal fluid that are involved in neurotransmitter synthesis and nerve growth. One recent study investigating Vitamin D and cognitive function found that the lower your Vitamin D levels, the more negative your performance is on mental tests. Another study found that people with lower vitamin D levels have slower ability to process information – especially in individuals older than 60. Vitamin D being a fat-soluble vitamin will need some good quality fats in the diet to be properly assimilated.

#2 – Fatty Acids

A substance called "arachidonic acid" is one of the most abundant fatty acids in the brain, and is crucial your neurological health, since it helps build the cell membranes in your hippocampus, helps protect your brain from free radical damage, and activates proteins that are responsible for growth and repair of neurons in your brain. In one study, 18-month-old infants who were given arachidonic acid supplements for 17 weeks showed significant improvements in intelligence, and in adults impaired arachidonic acid metabolism or insufficient arachidonic acid intake is linked to brain issues such as Alzheimer's and bipolar disorder.

Good food sources of arachidonic acid are Tilapia, catfish, yellowtail and mackerel, fatty cuts of meat, duck, eggs and dairy.

#3 – Good Coffee

100 mg of caffeine, close to the amount you'll get in a cup of black coffee, has been proven to improve memory recall. Caffeine's psychostimulatory effects are primarily because it blocks a receptor in your central nervous system that is responsible for binding a compound called adenosine. When you inhibit adenosine, you get increased activity of dopamine and glutamate, two feel-good, alertness-increasing brain-stimulating compounds.

However, more caffeine is not better, since higher doses may decrease blood flow to your brain, and you can quickly build up tolerance. Furthermore, you need to get it from arabica beans, and not coffee powders or substitutes, since cheap coffee and coffee knock-offs are high in mycotoxins, which can actually give you "fuzzy thinking."

#4 – Light Therapy

A dip in alertness and focus during the day can often be due to excessive melatonin, which can induce sleepiness.

It appears that the best way to increase mental acuity and focus during the day is to advance the melatonin cycle so that it finishes before you even wake up. Basically, you do this by limiting your exposure to blue light in the early evening – via both limiting use of TV's, phones, and computers at night, and also using blue-light blocking glasses, applications like Flux, and computer screen covers in the evening.

#5 – Fish Oil

Electrical signals used in thought, memory and processing bounce around in your brain and get transferred from one brain cell (neuron) to another via a point called a synapse, where the signals cross a physical channel before moving on to the next neuron. The walls that these signals need to pass through are comprised of cell membranes made up of about 20% essential fatty acids – like the Omega-3 fatty acids found in fish oil.

Specifically, these Omega-3 fatty acids may make the membrane that holds these channels more elastic, making it easier for the channels to change shape and for signals to propagate throughout your brain. With inadequate Omega-3 fatty acids, these channels lose flexibility and electrical impulses become hindered. It turns out that the use of Omega-3 fatty acids like fish oil may reduce severity of dyslexia and attention deficit hyperactivity disorder (ADHD), Alzheimer's, brain atrophy and cognitive decline, while simultaneously improving mental function.

#6 – Music

In addition to helping you exercise harder, music has been proven in studies to assist with "dopaminergic neurotransmission," which basically means it can cause a giant dopamine release in your brain, and make you smarter and more mentally responsive. Exposure to music also significantly increases blood flow, which, via something called a "calmodulin pathway," may cause a reduction in blood pressure and increased blood flow to the brain. While music can be distracting, light soothing music in the background can be useful for stress reduction and increase alertness.

#7 – Alpha-Lipoic Acid

Alpha-Lipoic Acid (also known as "ALA") is a fatty acid that can protect neurological decline with age, and can also be used as a treatment for diabetic neuropathy. Alpha lipoic acid can easily cross the blood-brain barrier (a wall of tiny vessels and structural cells that protect your brain) and pass easily into the brain to have these neuroprotective effects.

#8 – Acetyl-L-Carnitine

Acetyl-L-Carnitine plays a variety of roles within your brain, including synthesis and stabilization of cell membranes, regulation of neural genes and proteins, better function of the "mitochondria" (the energy powerhouse of the cell), protection from free radical damage to the brain, better transmission of acetylcholine, and enhanced glucose uptake to the brain.

In addition to increasing alertness, cell energy producing capacity, and neuronal transmission support, Acetyl-L-Carnitine can also help with depression symptoms.

#9 – Vitamin K2

Your brain contains one of the highest concentrations of vitamin K2 in your entire body, and it is in this area of your body that Vitamin K2 prevents free radical damage to neurons and contributes to the production of the protective "myelin" sheets around your brain cells.

Vitamin K2 is a relatively new darling on the supplement front, and many folks are rushing out to buy and use it for its bone building, brain building, and other remarkable benefits. You can get it from natural sources: grass-fed beef, fermented dairy products (like kefir) and natto (a fermented soybean derivative).

#10 – Brain Aerobics

There is continuing research that doing brain aerobic exercises (like Sudoku) can help to "age-proof" your brain and slow the onset of symptoms of brain aging, and can also help to keep your brain functioning at peak capacity. To qualify as a good brain aerobics exercise, an activity must have novelty, variety, and challenge.

In other words, going to work every day to your "mentally challenging" job does not provide the novelty to challenge your brain, sticking with the same "brain aerobics" activity day-after-day does not provide the variety, and engaging in brain activities that are familiar to you or easy (such as playing the same challenging computer game every day) eventually does not provide the challenge.

Finally, one of the best ways to get smarter is to exercise. But not all exercise qualifies. In fact, aerobic exercise is best for boosting brain power.

The ability to focus can make or break your success in life. In fact, scientists consider the ability to establish mental focus as an important predictor of a person's future success. It is perhaps one of the most important indicators, along with intelligence and GPA. We may be unable to increase intelligence, but we still can greatly increase our ability to focus. Having seen in the earlier chapter on how various factors affect the manner we see and sense the world due to the amazing variety of stimuli bombarding us, we can argue that retaining focus is the most difficult activity as we go about our daily lives. However, this art of focusing is an absolute need for not only survival, but for excelling in anything we do. An individual should train himself or herself to be self-aware of how good their quality of focus is, as they go about their jobs. This will remind themselves not to be victims of dangerous mind-wandering which may end up in unwanted consequences. Similarly, in teams, each member should be made aware to watch each other for signs of lack of focus or alertness as part of their routine activity. These kinds of trainings are further discussed in the chapter 11, "New training for cognitive risk awareness."

PART 3

Hitting the Nail on the Head

CHAPTER 10

A Holistic Safety Strategy

Safety is something that happens between your ears, not something you hold in your hands

— Jeff Cooper

DEVELOPING A HOLISTIC SAFETY STRATEGY

When the new seafarer is being screened for selection by the crewing manager, how much scrutiny is involved in ensuring that this new employee is a very good fit for the company and the job at hand? How do they do this evaluation? Is it merely by verifying documentary evidence of competence and past service records? And a medical fitness test that evaluates only the physical aspects of health? Or is it an independent method utilized for bias-free evaluation of the person to fit physically, mentally and culturally to the new organization? How many companies ensure that this is done at the basic stage of selection?

If any selection is completed without due diligence and checking these qualities in the new seafarer, let it be not surprising for the management if deviant behaviours are quickly being reported about the new employee. So, how many companies take care to complete these preliminary screening checks before employing the new employees? This is critical to prevent future accidents and losses. After all, any worker, even at the lowest level of the hierarchy is capable of setting off serious incidents or accidents. In crew manning offices, when the need to fill a position, say a senior rank officer, holding the right competency qualifications and experience is to be completed, there is a great deal of activity to be completed. After verifying several potential candidates, the one to be selected should not only have the right set of competency qualifications, but also the minimum number of years of service in the particular type of ships and trade (especially on tankers) and this experience has to be complementing the experience of the other officer staff on board the vessel. After this basic requirement, that officer should hold certain minimum training certifications over and above the competency requirements. This may involve doing additional value-added trainings for which courses will have to be arranged. After this, company approval for this officer will be required at higher levels of management. If the officer comes this far successfully, then a stringent medical test is to be completed successfully. By the time a crew manager finishes with these procedures, they may get the short end of the stick if that officer is snatched by another company that does not take as much time in the selection and screening process. So, a crew manning company has to keep a close watch on completion of the process if the officer is in a hurry to join a ship as his wages will start only after joining. This contract system for seafarers is the bane of the shipping industry as it sometimes causes the seafarers to act as heartless mercenaries who will jump a company to the next highest bidder, even if the increase in wages is only marginal. This system has never augured well for developing employee engagement and instilling company culture and value systems in seafarers unless they have sailed for many years on this contract system with the same company. That is sadly not the case always. Imagining oneself in the shoe of a crew manager, one can see how the anxiety to fill a position successfully is a tiresome effort and it won't be surprising if crew managers try and take short cuts in the overall screening of the joining staff. And to think that this joining process is happening for all ranks on a daily basis, one might wonder how many times the crew managers have completed the "letter" of the law, but not the "spirit" of the law in screening employees. How many potential accidents are set up when this screening process is compromised? This is a matter for ship operators to introspect and check if their processes are water tight and crew managers are sensitized to the extent of their responsibility in ensuring safe ship operations.

Without doubt, shipping companies should continue to follow the good practices that traditionally brought decent safety results without diluting any of the good processes involved. This means that checklists, permits, procedures, stop cards, incident reporting, safety workshops, resource management courses, people skills training and refresher trainings will continue to be used. Obviously, these are be carried out in tandem with new trainings that will be required for competency enhancement and skill trainings (especially for new equipment) including refresher for these activities at periodic intervals. Also, we must look at having a fresh approach to the traditional reporting we have used as its failures may be due to issues that were not properly recognized.

New trainings will continue to be required for all that new equipment under development and new regulations that will always be necessary as we advance in the world of technology for commercial expediency and regulatory compliance. Companies must slowly gear themselves for the new changes in the shipping industry that will revolutionize seafaring as we know it today. A broad effort is going on to bring down the dependence of manpower on the carriage of goods across the seas and coastal routes. So, a seafarer of tomorrow will be expected to play the role of a navigator engineer, supported by a plethora of new technology in real time from ashore to safely run, monitor and improve all the shipboard functions. Hence, it would be prudent for maritime administrations to look at new competency and certification requirements for the future seafarer, including the shore based remote monitoring/support teams that will take an active part in the success of these ventures. While these trainings may be targeted for operational competencies, safety will be affected if the trainings are done half-heartedly.

The path to unmanned or autonomous ships may take long and our concern for the present is quite important to maintain, as traditional shipping will continue to be the norm for most of the shipping industry, including coastal shipping for years to come. Eliminating human losses, property losses and environmental damages are what we need to focus on while we strive to safely move cargoes across oceans. Hence, the strategy for meeting these objectives will need to be innovative and practical in order to be accepted and succeed. Seafarers have a famous (or infamous) tendency to shudder at the thought of any new training. However, new ideas generated by research in various fields should not be overlooked that have a direct impact of how and why they exhibit behaviours as demonstrated that impact safety. Hence, we have to craft a new safety strategy that will improve dramatically the present understanding of root causes and without fail, will improve safety standards, thus leading to a more effective and sustainable safety culture.

After having seen how we had failed to incorporate our understanding of brain functions into our safety trainings, it is now high time that we closely study why in spite of huge investments in training, we still face the regular onslaught of accidents and near misses that are causes of grave concern. To understand this, we need to see how any seafarer perceives the world around him or her while carrying out any activity. The earlier chapters explaining the manner our brain senses and perceives the world clearly tells us that we have blindly ignored this powerful aspect of our reality. This means that the usual understanding of what we consider to be our true representation of the world around us may be flawed and this realization must be properly recognized. Are we really seeing a hazard with a clear eye or through a coloured prism? Do we realize how it diffuses our vision of the reality and therefore put us in the line of danger? Are we ready to fully grasp and understand these phenomena and incorporate them in our trainings? If so, we will start progressing in really understanding root causes of accidents and prevent these occurring proactively. What we call today as preventive actions are still not sufficiently proactive enough to qualify as actually preventive. We have shied away from attacking much more deeper issues that would have led us to the "root" causes of failures. We failed to look inwards at the source of all flawed thinking, i.e. our seat of thinking, the human brain.

So, for a holistic safety strategy to be deployed across a fleet, both at the management office level and shipboard level, there should be a complementary approach to prevent poor performance and create safety excellence. Some of the suggestions given below refer to the traditional ship management issues and others deal with what we can call the psychological weak links or cognitive barriers to thinking that are usually ignored. However, it should be borne in mind that a comprehensive safety strategy will encompass the best practices that great companies are already using in running their operations and planning which impacts their people, processes, assets and environment. Additionally, embracing behavioural, cognitive and social psychology paradigms will complete the full picture in crafting a holistic safety strategy.

Finally, following suggestions are given for management and shipboard levels to enhance some of those good practices where it is expected to provide good dividends with reference to safety and performance:

At Management Office Level:

1. Top management should be establishing clearly understood organizational values that will be invisibly running through transactions at all levels of the company and on-board vessels. These values and beliefs are the social 'norms' within an organisation and influence the way an individual act, when operating the social

context of that organisation. The culture conveys a sense of identity for employees and is believed to facilitate a sense of commitment and act as a mechanism to guide and shape behaviour. Top management should be visibly adhering to it in good times and bad sending out a strong signal that there is no safety culture without a strong organizational value system. It goes without saying that values of integrity, honesty and transparency are the backbone of any enlightened organization and this should be inculcated at all levels of the organization. In the long run, an organization diligently practicing these will notice that this also makes huge business sense as an entity and also bring enjoy greater employee engagement. No employee should join a vessel if he/she has not fully internalized the organizational values of the company. This training should be enforced at all levels, but very importantly should have top management's visible support and energetic direction to become successful.

2. Excellence organizations lead by values, more than they manage by rules. Very strong & visible commitment from management leadership that safety is absolutely the top priority in the running of all activities, both ashore and on-board ships. This should be visible at all events, seminars, and company gatherings where safety briefings should be clearly done at the start of all events (preferably by the top leadership). An evangelist style of spreading safety messages is a must! Something akin to the manner what CEO Jack Welch did for spreading Six Sigma in General Electric in the nineties. Without doubt, real commitment must begin with top management and be communicated effectively to other company leaders so that they aggressively find ways to establish, demonstrate and perpetuate a pledge to genuinely create a safe workplace.

3. A very careful consideration should be made when setting KPI's and targets for shipboard activities and

personnel, such that these do not assist in creating or setting up accidents and losses. (For e.g. KPI's that sets targets of outstanding PMS jobs in a month will end us either making the ship staff hurry with jobs to complete or tempt them to fudge figures thereby defeating the purpose of safe operations. Efficiency should not win over effectiveness.) In addition, zero based targets should be avoided as it will lead to more risk taking in the longer run. Goals and targets should be made so that it represents measurements showing how you want to achieve rather than what you want to achieve. Organizations that have achieved sustainability of excellent results in culture and performance define, measure, and motivate what they want, rather than what they don't. When we measure success by negatively reported outcomes, we are driving safety culture excellence by reaction, rather than proactive identification and response.

4. Similarly, performance targets laid down for the vessel Superintendents should also reflect realistic and safety focussed objectives as they should not push ship staff to focus on their own (Superintendent's) personal goals to the detriment of short- and long-term safety. Overall, from the top management to the ship staff level managers, the goals of the organization should be aligned to ensure common standards for safety and performance are maintained.

5. In incidents involving "human error" as the root cause, the investigation process should include steps upward into the decision-making that may have influenced what the worker did instead of blindly seeking modifications to worker behaviours. Accidents' causes are embedded in *why* people do what they do. They seek to discover and remedy the real reasons for poor performance, including leadership values, organizational processes,

organizational structure and management practices. A shift in belief from "people are the problem" to the recognition that "process is the solution." (1)

6. Management level representatives going on board for ship visits should adhere to the highest standards of safety and compliance with all regulatory requirements. They should also ensure that ship staff is in total compliance and assists them in obtaining compliance rather than punishing them for not achieving. Superintendents should clearly cultivate an image as a solutions person rather than as a problem creator. They should be able to seamlessly work with the shipboard teams with a "solutions" mindset to enhance overall productivity and be a compassionate and transparent partner to ensure their wellbeing. A propensity to react with blame games will push the ship staff to get inside their shells and this will severely affect transparency and trust. This will do wonders for the ship-shore team bonding and lead to better safety performance. Very importantly, these reps should uphold company values and beliefs without fail and be very stringent in ensuring that these are maintained at all levels.

7. Shipping is one of the most hazardous industry and we should not lose sight of the fact that our seafarers are operating under great dangers and a unique dynamic environment. Many ship operators are usually persons who were ex-seafarers themselves and have a better understanding of these conditions under which ships are operated in. However, not all commercial operators may have such an intimate knowledge of the real world and tend to focus on the commercial nature of the venture. The "profit-minded" approach is important for the commercial success of the company; however, this should always be secondary to the safety focus. Many a shipping operator have to their chagrin found this to be

true after suffering massive accidents, especially fire and oil pollution incidents. In some cases, the companies have had to shut down. Hence, a strong safety focus is always a sure recipe for long term success! If all ship operators had a "good ship owners" approach to running their ships, its equipment, machinery and people, then it would bring an ideal state of affairs with the long-term outlook embedded in the daily operations. However, this is not the case in the real world as many operators and ship managers are here for the short term and the severe competition and commercial interests ensure that the main focus of some of these companies is always survival in a cutthroat world. In such a scenario, it is difficult to ensure the safety goals are met, as while they may have their good intent to prevent unsafe incidents, it may not remain so in practice. Due to various reasons, especially lack of resources, many companies cannot do justice to their efforts to improve both productivity and safety. It would be prudent for ship operators to ensure sufficient budget allocations for performance downtime due to safety reasons and the strict implementation of a superior safety culture.

8. People selection in the various management offices should be done with great care and a good training program to sensitize them to the nature of seafaring and the special needs of seafarers must be in place. Special efforts must be made to seek cultural fit among teams after understanding various cultural dimensions as per Geert Hofstede's findings. Or as some good companies do, psychological assessments may be undertaken to obtain the right fit for the various teams in the ship manager's office.

9. Inside management offices, there should ideally be unrestricted flow of information about company matters across departments and individuals in a free and

transparent manner. There should never be a "need to know" attitude when sharing company information as these practices will slowly lead to silos forming among staff causing dis-affection and adversely affecting employee engagement. An atmosphere of transparency and trust will be created when employees understand that the management will keep them fully informed of all the important developments, whether good or bad, thus preventing rumours from doing their ugly rounds.

10. Bad office staff habits, for example, loose blameworthy talks and deprecatory statements made against seafarers and sometimes overheard by them during office visits usually end up back firing. These statements can have very damaging effects on the image of the office personnel and a false feeling that all of the management staff are thinking the same way. This feeling of "us" vs "them," if instilled, can then spread at an alarming rate.

In addition, it is important to understand that negative and positive conversations produce a chemical reaction in our bodies. During appraisals and feedback sessions with staff, it is important to understand the negative effects of cortisol producing conversations and positive effects of oxytocin producing conversations. However, cortisol is like a sustained release tablet, the effect of which can last for a long time if the person continues to think over the bad interaction. It also affects a person's ability to connect and think innovatively, empathetically, creatively and strategically with others. In contrast, oxytocin's positive effect is very short lived.

Managers should be mindful of their conversations with subordinate staff and seafarers as it does not augur well for safe operations by a demoralized workforce and it is the least fertile ground for good ideas and improvements. Not to mention safety performance improvements!

11. Crew manning procedures should be carefully assessed to ensure that all required processes are being completed properly so that the selection of ship staff is done with the highest rigour. Any crew manager being a victim of high anxiety due to "joining" deadlines are supported so as not to compromise the screening process. They should be sensitized to their contribution for the safety of people, vessel, cargo and environment and the critical role they play in ensuring it.

12. A good strategy should promote the desired state of the culture, the organization seeks by ensuring the right leadership style. If leadership is strongly command-and-control, a dependent culture will not evolve toward independence. Leaders must direct and support such a change by shifting themselves and their focus on leading the culture by example as well as by direction. The old saying, "When the leaders lead, the followers follow," applies strongly to culture change. The next step is sharing the rationale for improving safety to create buy-in and strong emotional attachment. When that step is achieved, the next should be to create involvement opportunities for workers to get some hands-on participation in safety efforts. Once the majority are participating, the next step is to create ownership for safety. Deming said, "People support what they create." Giving creative input is one way to create ownership, but not the only one. Creative leaders have found many organization-specific paths to creating a greater sense of ownership in safety efforts. A key to getting engagement to the ownership level is to make the goals of safety people-related rather than focusing on numeric goals. Ownership is about the heart, not the head.

13. Safety professionals are leaders also and need to evolve their roles to facilitate cultural change. When workers are just learning about safety, safety professionals must direct their efforts. When workers begin to take initiative,

safety folks must change from directors, to providers, to supporters, to expert information resources. Sadly, most safety departments are understaffed and the people staffing them get caught up in reactive activities and never evolve themselves, much less the culture. When safety managers are constantly putting out fires, they don't think strategically about how to prevent them. Many safety departments need the same progress—from dependent to independent—that they wish on their cultures.

14. Eliminate all financial based incentives for shipboard staff when talking of safety performance. These usually have the least motivational effect. After all, they have legally contracted to work with the highest competency and safety for the agreed wages. Any idea about bonuses for good performances, if implied in the contract, makes the seafarer feel that it is an entitlement that they should rightfully be paid (after all, they are aware of Cost to Company-CTC). And if due to some reason, it is withheld, there is no better recipe for dis-enchantment and poor engagement. "Expectation management" is easier if unnecessary expectations are not created. Rewards and recognition for good safety performance (not operational) could be introduced in a fair manner so that teams are given awards encouraging better teamwork. These results should be declared periodically and publicized to create high visibility and importance. When such awards are to be presented, members of top management should make effort to be part of these ceremonies as it sends a powerful message of safety.

15. Also, eliminate all punishment-based targets as these propagate a "fear" environment and do not allow natural creativity and willingness to develop and achieve targets.

16. In all Company gatherings/seminars, highlight the ones who are the "positive change agents." Spread their message

and methods. Such a culture of innovation will improve creative thinking and performance enhancement.

17. Safety policies, instructions and processes should be also communicated in the language of the seafarers apart from English. This should be done even if it is understood that English may be the common communication language on board. There could be those few who never came forth and expressed their inability to understand instructions in English. In addition, use of non-verbal instructions will also assist in imparting knowledge for those who are visual and kinaesthetic learners.

18. While high crew retention rates are quantified and treated as a major achievement in the running of vessels, another important metric to introduce is retaining the members of good teams for the same vessels as it is well known that high performing teams will give better safety performance. All efforts should be made to make the same people return to the same vessels or type of vessels with nearly the same teammates. Deep expertise will be developed and retained which can be exploited in the face of pressure and which permits them to read complex situations, project into the future, and to follow timely and effective courses of action. This can be beneficial in the long run for both safety and performance.

19. Decision making must be based on systems thinking. All safety-critical industries are formed of different organisations which must interface successfully. In the maritime industry, these include shipbuilders, shipowners and managers, Masters and crews, port authorities, flags, insurance clubs and so on. In the absence of applied systems thinking, organisational decisions are taken that are locally optimised (i.e. too narrowly focused on a small part of the problem) at the expense of global effectiveness. There are countless examples of this in

the maritime industry – mostly driven by apparent opportunities to save money in the immediate future.

20. To enhance engagement of employees in safety efforts, clear definition of what that engagement looks like or what specific measures to adopt to make it happen should be made. Depending on the current state of the organizational culture, engagement strategies should first target creating a sense of belonging: a pride in the organization and a sense of team.

21. Coming to various trainings that both shore staff and ship staff undergo, a new strategy should be prepared to ensure that all trainings follow a two-pronged approach. An external factors' approach to teach the main content (theoretical and/or practical) aspects of the topic which may include the regulatory, knowledge and skill aspects of the training. However, this should be supplemented by the internal factors (or Cognitive aspects) which will impart those skills to the trainee to understand how he or she should be prepared mentally to undertake these activities in a safe and effective manner. This part of the training will have to be individually designed as per the topic and location of activity so that the participants can fully internalize the external aspects of the training by vividly imagining the areas of critical interest and how to maintain focus and alertness during the full activity. It would be important to include the thoughts, values and belief systems of the organization into each training so that every training is properly integrated with the organizational culture and not separate from it. Also, these are highly influential in impacting the social behaviours (including safe behaviours) of the trainees (both at shore and sea).

22. For all important trainings carried out for staff, a regular visible presence of senior management should be and there to demonstrate the importance of the training and

their commitment to safety. This connects the trainees to the cultural, organizational and managerial factors which will be influential to bring out the right behaviours while on board.

At the Shipboard Level

1. Clear understanding and implementation of the Organizational values and strong drive for a safety culture at all levels on board. No exceptions allowed at any time. Senior officers should be the embodiment of these values and reflect it at all times.

2. Like the regular shipboard safety drills, training sessions to be conducted for all crew on how to perceive the reality or likely distortions in reality around them and how it can differ from person to person. Also, that risk is a subjective variable and not an absolute idea is to be understood. Even if the Risk Assessment (RA) exercise is done and the risk quantified, it still does not make it an absolute value for ever. This training should be made mandatory until the concepts of risk and its perception in all day-to-day activities becomes easily understood and applied.

3. An understanding of brain functions will lead to awareness of what is peak and non-peak state of the mind. This information should be used for selecting teams so that apart from the concept of "rest' (which is not fully understood), we have other methods of identifying safe team members. Also, how to keep highly alert and competent members of teams working safely by monitoring certain physiological characteristics should be understood.

4. A simple understanding of how diet plays a great role in brain chemicals activity or inactivity is to be imparted. Good decision making involves clarity of thinking and ease of recall. Experiments have shown that the brain

needs essential fats for its sustenance and good fats and oils should be part of the daily diet of the seafarers. Also, a good diet has its beneficial effect on the microbiota can lead to better performance if the gut bacteria is kept in a healthy balance. This may need a substitution of shipboard diet with a healthier option and ship staff should be encouraged to eat healthy. These will make changes to the brain chemicals that will enhance alertness while on duty and keep the willpower to perform safely at a strong level. For example, lack of sufficient brain fuel (glucose) can make workers give up their attention and focus earlier than if properly fed, sometimes making them prone to taking unsafe decisions. In addition, it is important that good diets recognize the critical influence our neurotransmitters have on our thinking and behaviours. These are made from amino acids, vitamins and minerals which are available only from good quality sources of food. The reason we have low neurotransmitters is due to low supplies of high-quality protein, vitamins, and minerals. Amino acids come from properly digested protein in our diet. The author recommends that professional guidance should be sought to design diets for seafarers that will nourish their body and the brain. These will go a long way in preventing lapses in thinking and unsafe behaviours.

5. Introducing steps to improve performance and safety by mitigating the deleterious effects of poor lighting systems, low but continuous background noise and low-level vibrations that gives rise to chronic fatigue and stress effects. In addition, the use of blue light filters may be considered on computer screens to avoid its adverse effect on melatonin production which affects good quality sleep.

6. Ensuring that ambient temperatures, air quality and ventilation inside a ship are properly monitored to ensure that performance is not hindered due to fatigue

and stress induced by less than optimum levels of these parameters.

7. If the Safety Management System is a complex set of documents, it would bode well to digitize and make it available for staff to utilize it as per the area of activity and without having to read through reams of pages to complete an activity. Checklists and permits should be made available in digital/online mode so that these are completed efficiently and does not become a disliked chore.

8. Create social groups and activities that will bring all crew physically together on a regular basis. This is necessary to reduce onboard stress related issues and improve engagement. In addition, crew meeting up physically is very important so that people are comfortable in their personal spaces being shared especially so when emergencies need to be tackled together and teamwork is essential.

9. Social media devices should be banned from all critical work areas during any critical activity. For example, watch keeping officers should not be actively using any smartphones or tablets while on duty as social media addiction is nothing short of a disease and will adversely affect safety. The large tech companies are at each other's throat to mine "human attention" as it becomes the new global asset, probably surpassing many traditional valuable assets in value. In this cut throat competition, many minds are falling prey to addiction for social media devices and it becomes difficult to escape the clutches of their firm grip. Seafarers will have to be sensitized to this new danger on their horizon and learn how to detox themselves. These behaviours will otherwise be the cause of new accidents as internet's reach is spreading to the distant oceans. Not only is excessive smartphone usage linked to distraction, but also leads to poor task performance as per new studies by researchers at Rutgers University (USA).

10. Management should ensure that a strong positive safety climate is sustained on board vessels and workers are empowered to have control of their safety and risk. A strong safety climate encourages staff to become more responsible for their own safety performance. In contrast, a poor safety climate encourages workers to attribute responsibility for safety to the company. **(2)** This research also determined that a greater sense of control over hazards increases workers' confidence in managing their interactions with workplace hazards.

11. Instructions given to staff during various trainings should be with a positive affirmation. Positive reinforcements to be used more than negative statements to ensure the limbic system does not enter into fight/flight response. The overuse of "Don't do this" is to be avoided as it is a negative affirmation and replaced with the positive affirmation "Do this..." This should be used wherever possible. For example, "Don't bend your back while lifting heavy weights" should be replaced by "Keep your back straight while lifting heavy weights." The nature of our brain to seek taboo pleasure (read risky behaviour) is strong and such negative lessons are better remembered than the positive one. Hence, a "don't do." instruction is what the brain seeks to try out as the pleasure of doing the wrong is deeply rewarding. All safety training content can be looked into to check for such usages and corrected accordingly.

THE COMPLETE PICTURE: A MODEL OF HOLISTIC SAFETY: (3)

Holistic safety integrates the behavioural, cognitive and social psychology paradigms at a philosophical level and applies them in an effort to both explain and influence human behaviour in the context of workplace safety. The proposed model, which includes the three components presents a more holistic approach to ensure employees are more likely to act safely. This integration of seemingly disparate concepts and constructs leads to a powerful whole that is greater than the sum of its parts.

An approach that builds on psychologically driven safety management interventions that are likely to drive safety improvement over the coming decades.

This takes the form of an in-depth training experience aimed at encouraging employees to choose to be safe; to acknowledge and assess their own safety and the safety of those around them. Put simply, the premise of Cognitive Psychology is that much of what influences our behaviour occurs 'below the surface' in our mental processing. Although behaviours and emotions can be readily observed, there are a number of components that interact to give rise to these behaviours.

The most effective strategies aimed at workplace safety include some of the traditional approaches as well as other strategies that target the internal processes influencing safety behaviours and outcomes. This is the fundamental philosophy behind developing new safety-based training programs that incorporate both cognitive and social factors.

This model suggests that learning mechanisms (traditional approaches including BBS) and social factors (such as the culture, norms and the relative effect of leadership), combine to influence cognitions (our thoughts, beliefs and values). Addressing these unobservable components, in collaboration with a more traditional approaches, can assist in ensuring workplace safety is more effectively managed.

CHAPTER 11

New Training for Cognitive Risk Awareness

"Education is not the learning of facts, it's rather the training of the mind to think."

– Albert Einstein

When we visit a five-star restaurant, what is our expectation? As our order is being placed, we notice a tiny speck of mold on the bread. Will we accept this quietly or jump up with disgust and loud protestations? When we buy a new premium brand car, will it be ok if the door lock has a minor issue? Why don't we accept these small deviations? Why don't we become more understanding of the fact that some tired Chef or a tired car factory worker could have easily made these minor mistakes? After all, they are also humans like us.

So, the work that a seafarer does in this high-risk environment calls for the five-star standard (or maybe more) of safety and work delivery too as the consequences are very high. Then why is it that we look at safety as something that is not part of us, but an external demand thrust upon us? It is time to change and look at every action going into each activity as an action to be done with perfection. Because, these actions define each one of us and finally the organization as a whole!

The reader who has diligently gone through each of the previous chapters will now have an appreciation of the intrinsic factors that shape and influence our understanding of the world around us as we grapple to get an accurate perception of the reality in which we must function. This reality appears to be as true to us as the seemingly near perfect replication of our image in a mirror. But mirrors can also obfuscate the image in many ways. Similarly, our senses and our brain can colour the external world and give it a different hue giving rise to an apparent reality that we believe as real and react to.

Hence, we now must peek in and see what those inner train of thought processes are as we get engaged in any activity, be it as an individual or as part of teams. In short, we must start practicing constant "mindfulness" in every activity that we commit to. Regular practice of this exercise will make the person automatically sense the real risks as it arises and the proper intervention by controls will be that much easier. In this way, setting intent and controlling the risks will be an inherent part of the activity and result in safer outcomes and happier teams. If after years of safety and competency trainings, we have still not been able to achieve the tag of a "highly safe mode of transport" (even if we are a highly economical mode of transport), it is high time we also go the extra mile and look at how to correct the seed of unsafe thoughts which have been not addressed in the various trainings designed for seafarers and managers. This can be achieved only by allowing us to delve into our minds and senses for which the right training needs to be designed and imparted.

In the below sections, some training schemes have been sketched, but these are only a guide and obviously not the only manner these trainings can be imparted. The objective should be to achieve the outcome of obtaining better awareness of team and individual mental states while at work. These trainings in mindfulness can be changed to suit the work processes and location of activity and as per the organization's core beliefs and values. These trainings can be used as a guide to prepare full-fledged, topic-based course contents and lesson plans and these may be changed and improved as required.

The broad objectives of these types of trainings are that the trainee should be able to:

- Prepare his mind for setting intent, become aware of hazards and risks,
- Mentally "see through" the whole process for safe execution of the activity,
- Recognize and watch out for his/her teammates for signs of lack of knowledge/skills, unsafe behaviours, poor focus, poor alertness, fatigue, other risk behaviours.
- Look inwards for any issues arising due to lack of mental focus, alertness, concentration, fatigue, brain fog, risk taking tendency, boredom, distraction or other mental issues.
- Visualize vividly safe & positive outcomes in each activity.

TRAINING EXERCISE 1 (SAFETY YOGA EXERCISES)

Without meaning in any way to disrespect the great traditions and ancient knowledge of Yoga, I am taking the liberty to use this term for this kind of exercise. However, seeing the similarity of the below exercise to many yoga relaxation techniques, I am tempted to call it this way. I apologize in advance to diehard practitioners of traditional yogic streams if this should in any way offend the them or any reader. But the idea is to make this practice easy to follow and improve safety, wherever this may be tried. In the end, the objective is to make the practitioner feel relaxed, fully aware and mindful of the impending activity and to be able to successfully and safely complete it.

The team leader should conduct this exercise with all the team members.

Once the "toolbox" meeting is concluded, the team stands around the team leader in a relaxed manner, hands at the sides (sitting would be ideal if it is convenient) and eyes closed. The team leader will take the team through the below mental visualization exercises:

- Take a deep breath and exhale.
- Repeat with another deep breath and exhale. Focus on the breath.
- Imagine each worker wearing their proper PPE (start from top to down).
- Imagine each PPE is in good working condition & properly secured (Visualize each one being used and take them through each equipment)
- Continue taking deep breathing and exhale. Focus on the breath.
- Imagine the location of the actual work area and making entry into space. Imagine each step of entry and imagine safely entering the space.
- Imagine all loose objects (ladders, manholes, staging, tools) secured correctly in position.
- Imagine taking up tools of work in your hands (think of each equipment or tool in use). Imagine their correct state or condition.
- Think of the environment required to work safely. (secured for weather effects, no presence of toxic gases, proper O_2 content, air temp is suitable, air ventilation is adequate, Lighting is adequate.)
- Communication is adequate and verified.

- Imagine you watching out and being alert for any new hazards that may arise.
- Continue deep breathing and exhale. Focus on the breath.
- Imagine doing the work activity assigned safely. Imagine being very alert and fully concentrated in work.
- Imagine team all working in unison and good cooperation.
- Imagine continuous monitoring of environment, process, people and equipment.
- Imagine getting successful result in your work. (Imagine final output expected)
- Imagine concluding work successfully and withdrawing successfully all tools, equipment, people.
- Imagine the feeling of successful completion. Imagine the happiness of being safe.

TRAINING EXERCISE 2 (BEING A VICTIM)

a. Visualize mindfulness of safety around us… Now imagine what it is to be fully safe in the work area. Now imagine you are victim of an accident…your leg is now badly injured and bleeding…… You are lifted off and put into a hospital. You wake up with a badly injured bandaged leg. Imagine all those things that went wrong and could have been corrected for preventing this happening.

b. Imagine work equipment is not fully in order. But work continues…soon it creates an accident as it breaks and a part shoots onto the worker's body piercing it. Imagine all those things that went wrong and could have been corrected for preventing this happening.

c. Imagine one other team member starts working recklessly and does not use safe PPE. You may be affected adversely by his actions and you ask him to correct the situation.

However, soon you suffer a fall to the ground due to his actions and get injured. Imagine all those things that went wrong and could have been corrected for preventing this happening.

d. Imagine you enter a workspace and commence working without checking the conditions of the space. A loosely secured item from the top comes crashing down nearly missing you. Imagine all those things that went wrong and could have been corrected for preventing this happening.

e. Imagine you are under a strict deadline and team members starts to hurry and work faster, occasionally taking shortcuts. Soon, one of them bypasses a step of checking the safety gear. Soon, there is an explosion and confusion all around. Imagine all those things that went wrong and could have been corrected for preventing this happening.

More scenarios may be conjured up for different types of work activities.

TRAINING EXERCISE 3 (SHARPENING KEEN FOCUS)

Concentrate on any one specific item of interest. Keep the focus on it for as long as you can. Keep increasing the interval of focus. Practice this regularly. (An easy way to start this is by focusing on one's breath for as long as one can).

TRAINING EXERCISE 4 (PERCEPTUAL AWARENESS)

From the earlier Chapter 9, "Focusing the Brain," we know that while the mind wanders, our sensory systems shut down, and, conversely, while we focus on the here and now, the neural circuits for mind wandering go dim. At the neural level, mind wandering and perceptual awareness tend to inhibit each other;

It is important to be in the "here and now" while carrying out duties that require a good degree of alertness. This exercise to stop mind wandering can be highly valuable to keep the work activity safe and efficient.

Each person to sit comfortably with closed eyes. Take a deep breath and exhale. Continue breathing and exhaling.

Become aware of the warmth of your breath.

Become aware of the pressures of the clothes on you.

Become aware of the pressure of your bottom on the ground.

Become aware of sounds around you.

Imagine the work area and the various hazards. Mentally see the various hazards and their likely consequences.

Imagine laying out required controls to each hazard to mitigate risk.

Imagine successful outcomes.

TRAINING EXERCISE 5 (SETTING INTENT)

Here is an easy process for setting your intent around certain goals and your future:

1. Think of the goal or situation that you would like to set your intent for.

2. Set intent for yourself in terms of the experience that you want to have in that situation, or in achieving your goal.

3. If there are other people involved, then set your intent for the kind of interaction that you would like to have with them. Perhaps you would like to have fun, learn something new, be productive, feel peaceful, be happy or loving, feel respected, be calm and helpful, or feel connected with others.

4. Create a mental movie of what you will be like in that optimal, future situation. Notice what you are experiencing in the situation once you have set your intent. What are you hearing? What are you saying to yourself? What are you seeing and what are you feeling?

How we think, act, and behave has a direct influence on the greater system of our external reality. When we set our intent, we are influencing both our inner reality, and our outer reality in a way that sets a chain of events into motion. We are bringing forth a new chain of events that are directly related to our deeper subconscious thinking, as well as our overall intent for the desired outcome and journey that unfolds.

In every Toolbox meeting, what is it that we need to focus on? Is it just the hazards and risk management that each member or the team will do? Or will we take a more powerful process together by focusing on the intent of this activity? So, make each member of the team imagine the outcome of the work at hand, the safe and desired outcome of the work that we need to call as a success. This intent focus is equally powerful as each team member will now be keenly sensing all those stimuli that will lead to the successful execution of the various duties.

As mentioned in an earlier chapter, the RAS (Reticular Activating System) is a powerful filter of thoughts and stimuli that affect the way we see the external world. Also, this affects to a large extent how we perceive the risks around us. For example, if we are standing in a crowded party amidst the din of voices and music, it is not surprising that our name being spoken in the crowded room catches our attention even though we cannot make out most of the sounds emanating in this room. Our brain is finely tuned to pick up such signals that matter closely to us. Another example is when we do research to purchase a new car model, we tend to be quite surprised when we start noticing the same vehicle model on the roads more frequently than we did earlier. This may sound like a coincidence; however, the brain is now tuned to focus on these images better. That shows how powerful this filtering mechanism is.

For safety, we must continuously train our minds to focus on safety aspects so that these risk signals are quickly noticed and brought up to our conscious mind for action. When we focus of safety outcomes, we quickly perceive when there are deviations from the planned outcomes.

And to make this a more powerful exercise, we can make the team members carry out a deep risk awareness session wherein each member will talk about his or her mental state and readiness to carry out the job at hand. They can also talk about how they feel at this stage. This will be keeping in mind all the lessons we have learnt from the previous chapters of the various influencers on the mind. The safety team supervisor will carefully assess each report with a keen ear for warning signs of deep mind risks. This may pull up newer risks that may not be identified while carrying out the formal risk assessment or toolbox meetings.

To gain an experience with setting your intent and positively programming your RAS, try saying the following three sentences to yourself:

1. *"I hope to enjoy my dinner tonight." (Notice how you think about this – your internal pictures, voices, and feelings.)*

2. *"I want to enjoy dinner tonight." (Notice how you actually think about this – your internal pictures, voices, and feelings—what is different from the first question?)*

3. *"I intend to enjoy my dinner tonight." (Notice how you actually think about this—your internal pictures, voices, and feelings—what is different from the first two questions?)*

Pay attention to how each of these simple changes in your language creates a very different experience. For most people, the first question will produce some doubt. In other words, multiple images will appear in your mind representing different possibilities—one is that you may enjoy dinner and the other one being that you won't.

The second sentence should produce a different representation. When you say, "I want to enjoy dinner tonight," you will typically see what you want in the future, but you may not see yourself having it now. The future then may feel compelling because you see what you want. But there is still some room for doubt because it is more difficult to put yourself into the actual experience of achieving it.

In addition to safety trainings that are done to deal with external threats and risks, the following list of trainings are included for assisting the seafarer to reflect inwards. This is another area for the origin of cognitive risks that are not covered in our usual maritime trainings. Also, these trainings remind each seafarer that in any work activity, it should never be taken for granted that only fixed risks exist, and that one risk assessment (RA) exercise prior to the work is fully sufficient to take care of the risks involved. This complacency is dangerous and has been proved disastrous many a time when it results in accidents even after following the "right processes"! This is a dynamic world of evolving risks and a major area of this origin is in our minds that we need to be watching very keenly and arresting those risks that are recognized. This training will remind the seafarer to become aware of these intrinsic risks.

However, the list of training topics mentioned below is not the complete list or a full training methodology. Every company can tailor their own structured trainings as per their needs and situations on board their vessels. This list is just a sample of trainings that may be used. A few of them are explained in greater detail for giving guidance to trainers so that they may develop newer topics of interest. They are encouraged to creatively make the situations more realistic so that the learnings are maximized and sustained. In addition, some more topics pertinent to understanding cognitive risks are given that may be added to the repertoire of trainings.

Knowing the high level of aversion to new trainings by seafarers, I should hasten to add that these trainings should ideally not be given as a separate module, but incorporated into the trainings of the regular modules. The intention is that the trainee always has a focus on these cognitive risks that were earlier not part of the regular training. Some of the trainings are better suited to be carried out on board the vessel or work area.

ADDITIONAL TRAININGS TO IMPROVE COGNITIVE AWARENESS

1. Trainings to increase power of observation for early identification of hazards.

Training objective: To increase hazard and risk observation powers		
Scenario: Team about to commence work in an enclosed space for hot work activity		
Description (5 minutes duration)	Pause and assess/verify all hazards in work area/arrangements/tools as per Toolbox meeting Discuss deviations and issues New hazards to be listed Proceed with work if no Non-conformity	Each person will mentally assess hazards in work area (360 degree approach) and allow as much creativity in identifying hazards. If no deviations, discuss possible changes likely. If serious deviation noted, consider aborting activity. Continuous assessment activity to be done
Training outcome	Trainees will be able to list all pertinent hazards in work area and identify risk from all sources of hazards.	
Remarks	This exercise can be replicated for all other kinds of work areas including work aloft, work over side, hot work etc.	

2. Training to recognize and be aware of continual depletion of will power under stress & subsequent possibility of unsafe decision making.

Training objective: To recognize effect of continual stress on will power to act safely		
Scenario: On a simulator, team issued with 3 to 4 stressful decision-making situations. Observe how it affects the person's mental stamina to take safe decisions or if indifference developing. Other team members to observe and sense in team leader.		
Description (5 minutes duration) Team objective for the exercise or work activity (Navigation or Machinery repairs etc.) is explained.	Work situations and minor emergencies set up in simulator with the team making decisions for safety and performance. Observing team to watch conflict arising between efficiency vs thoroughness, fatigue onset with each stressful decision, Loss of situational awareness, shortcut taking, confused reactions etc. End exercise with a light carbohydrate snack to recover any loss of will power to aid in debrief. If required, follow up with one more stressful situation to demonstrate increased alertness/ will power after carbohydrate snack.	Navigational situations in high traffic areas or Engine room trouble-shooting exercises with ship operating in pilotage waters. Debrief with team leader explaining mental states at each decision taken and effect on safety.

Continued...

Training outcome	Trainees will be able to spot changes in mental states of co-team workers/leaders from changes in behaviour and mental alertness.
Remarks	Repeat exercises with other work scenarios and add new decision-making situations to test the ability to continue the mental focus for safe decision making.

Alternatively, the above exercise can be done in a classroom setting by asking the trainees to solve a puzzle (or a nautical/engineering/seamanship problem). Once they complete it, another puzzle (or shipboard issue) is given to solve. It will be explained that this is the last problem of the day. Once they complete it, yet another problem is given to elicit dismay or frustration. Once they complete it, a group discussion will be done wherein, each person will explain what they thought or felt at the end of each exercise and what they underwent when the problems kept coming. They should be requested to give explanations to what they felt and finally, the trainer should connect these thoughts and feelings to real shipboard situations when similar situations are experienced. This should be followed by explaining what the cognitive basis is and how the brain is struggling to get its fuel for the problem solving. The training can end with a light carbohydrate snack and the trainees can sense any relief from the negative feelings earlier experienced.

It is to be understood that the experience of each person may be different from one another. Some can accept more stressful situations and handle it better than others.

3. Observing team worker responses:

Training objective: To remember to pause intermittently and observe co-worker behaviours and recognize unsafe tendencies
Scenario: Overhauling of piston in Engine room or similar team activity (Repairing crane motor on deck)

Description (3 minutes duration)	Brief each team member to observe other team members for various external signs of cognitive risks and mental alertness. Each member to pause work at regular intervals and watch another member to ensure new risks are not arising during activity. Team leader to pause and take status check to agree and continue with work. Proceed with work if no Non-conformity	Each person will mentally observe other team members for, focus, fatigue symptoms, yawning, weakness, poor posture (slumped), distraction, multi-tasking, anxiety to complete, wrong use of tools, etc. If no deviations, discuss possible changes likely. If serious deviation noted, consider aborting activity. Continuous assessment activity to be done
Training outcome	The trainees will understand the importance of the habit of frequently watching out for any unusual symptoms of human cognitive risks exhibited by team workers.	
Remarks	This exercise is mainly to inculcate the habit of watching others for new risks. The need to occasionally pause one's work to deliberately watch the other team members is to be stressed. The focus of this exercise should be on creating this habit than dwelling on what is observed.	

4. Understanding signs of alertness/lack of alertness/fatigue symptom

Training objective: To understand the various signs of lack of alertness and fatigue exhibited by team worker
Scenario: Navigation in poor visibility conditions in English Channel

Continued...

Description (3 minutes duration)	Brief each team member individually and instruct each to exhibit certain specific symptoms (fatigue, lack of alertness, foggy brain, forgetfulness, slurred speech etc). Other team members to pick up such signs each time they observe or state how many times they noticed these arising and from whom. Each person not to know what was briefed to others. Team leader allowed to rotate duties of some team member if symptoms persists with any person.	Each person will mentally observe other team members for, focus, fatigue symptoms, yawning, weakness, poor posture (slumped), distraction, excess anxiety, panic, wrong commands etc. Debrief and discuss the observations recorded by each member and not reported by the team.
Training outcome	The trainees will be able to recognize various types of symptoms indicative of human cognitive risks exhibited by team workers and take corrective actions.	
Remarks	The team may be asked to discuss other kinds of behaviours they have experienced indicating similar risks.	

Suggested new training topics for a good understanding of the pitfalls of flawed thinking giving rise to cognitive risks at the workplace:

- **Improving Situational Awareness:** Exercise to constantly focus on situation now and the upcoming one with a keen eye on the risks involved.
- **Bias Understanding:** Preventing assumptions leading to bias that attractive person or senior rank person as more

smart/intelligent needing less instruction/monitoring. More exercises can be designed for other cognitive biases.

- **Perception of Reality:** Realization that the reality out there (especially the work place) is only as true as our senses and brain allows us to see. It is never complete and absolute. Teams to factor this risk in all work activities by using a questioning attitude to non-fact-based assumptions.

- **Balancing Risk Perception after a Major Incident:** Shop attendants at a perfume store make one smell coffee powder after the customer is exposed to a particular perfume. This removes the memory of the earlier scent and prepares our nostrils for the next perfume trial. Otherwise, it will be difficult to feel the real smell of the newly offered perfume. Similarly, a safety risk perception refreshing is to be done after any high-risk incident or experience in order that the previous untoward incident does not dull our senses to the risks in forthcoming work activities. Anyone who have experienced high risk situations may tend to lower their guard when exposed to lower levels of risk, which can however lead to incidents. This should be guarded against.

- **Panic and Mental Hijack (of Right Brain):** When an external stimulus is interpreted as a severe threat, a panic situation develops. We are mentally hijacked as the amygdala circuitry captures the right side & takes over the bottom brain which then drives the top brain. But the left pre-frontal side can send signals downward that calm the hijack. Active engagement of attention signifies top-down activity. This focused goal-oriented attention inhibits mindless mental habits. In short, becoming mindful of the present can rapidly bring back control to the top brain. Training in "mindfulness" and its ability to calm frayed nerves is very useful in many situations.

- **Memory and Recall:** Human memory capability is limited and too many detailed instructions cannot be followed due to memory recall problems. Break down complex instructions to simple sets or smaller chunks of information. This can avoid wrong actions due forgetfulness. Too many instructions in one go can cause a mental overload and resultant stress. Don't exceed more than 4 chunks of instructions at one go. That is the average limit of human capacity to hold in memory.

- **Risky Decision Making/Bravado Styles of Expression:** Team members to observe and flag off such behaviour and they should be capable of observing oneself and others in active situations.

- **Avoiding Hurry and Anxiety Driven Behaviours:** To recognize anxiety ridden behaviours and tendency for mindless hurry can itself be a huge risk driver. To "slow down" when such situations are recognized as developing.

- **Mental Frustrations and Disenchantments at Work:** Recognizing these negative thought processes in oneself and others and learning how to cope with these by having constructive and sympathetic "conversations." Disenchanted workers are a great risk to themselves and to others. Training and awareness in having difficult conversations to release all negative thoughts and resolve these pent-up issues.

- **Distraction Effects (Colleague/Family/Friend Related/ Money/Social Media):** Mind wandering or distraction should be tackled with exercises in "mindfulness: Distraction effects can be very severe that the worker may not be able to recall why things went wrong even though it happened in front of him/her. The Bottom up brain can hijack automatic responses (need to look into social media at frequent intervals etc) and bring down situational awareness drastically. "Mind-wandering" is considered as a default mode of the brain and learning to retain focus has to be constantly practiced.

- **Deadline Anxiety and Risk:** If deadlines are not realistic, to re-assess and schedule work activity to achieve safe outcomes rather than create high anxiety and unsafe acts. Mindfulness trainings to bring down anxiety attacks and re-assess workload to make it more achievable.

- **Multi-tasking Beyond Ability:** To understand that excessive multi-tasking is the setting up of a new risk as it can create high stress and a tired, depleted brain. To learn to prioritize work and share duties to ensure that multi-tasking is kept to practical levels.

- **Poor Teamwork/Low Trust in Colleague/One-Man Team:** Lack of trust can lead to acrimonious relations among team workers and undue stress in each person. This itself can give rise to stress reactions and poor cognitive processes leading to high risk situations.

- **Over Trusting or Fawning Behaviours with Seniors/Experts/Gurus:** When seniors or experts speak, juniors or subordinates should ensure that they have to maintain their critical reasoning and the ability to question (being assertive) at all times. The tendency to fall into passive submission mentally and not questioning when in doubt is a sure sign of future risk in the making. Seniors also to be trained to watch out for these behaviours to ensure that instructions are properly understood and followed. Always, the ultimate goal of safe performance should be kept in mind and not ego satisfaction.

- **Exhibiting Thrill Seeking (risk behaviour) Due Recent Boredom/Inactivity:** Proneness to seek short cuts, higher speeds, adrenalin rushes, jumping off heights etc after recent inactivity or state of boredom is to be observed and curtailed. To understand that risk-seeking is part and parcel of the brain's need for excitement and variety and conscious control is required to prevent it.

- **Over Reliance and Blind Belief in High Technology/ Computer Software:** Healthy distrust for high tech equipment outputs or computer-generated results should be maintained and observe in oneself the mental weakness of not challenging these results on a periodical basis.

- **Sunk Cost Effects:** The propensity in a team leader to add more resources to an already expensive project even though the results are below par and not desirable. Fear of loss of face/reputation makes the team leader invest more into the failed project to postpone the ultimate failure. All because of the huge investment already sunk into it. To recognize such situations and cut loss/damage in shipboard situations.

- **Confirmation Bias:** A confirmation bias is a type of cognitive bias in which information that confirms previously existing beliefs or biases are given preference. This can prevent a team member from using important information just because it did not agree with his/her previous understanding or opinion of the information. Eg. If the person worked with one reliable model of gas meter in another company and now sees the same model in the new company, however the equipment has not been serviced as required. He is likely to trust the readings due to the previous experience. This bias must be recognized if it has developed in any team member.

Once a workshop or training is provided to all safety players on the above facets of cognitive characteristics, a training in how to complete a Risk assessment to include these kinds of risks should be undertaken. This will entail a separate RA exercise for only the above cognitive hazards if applicable to the job at hand. This is specifically recommended for all those critical jobs where the losses or impacts may be significant and should be done prior every critical activity for which checklists and permits are also used. Also, during the toolbox meeting, a special session on cognitive alertness should be completed to ensure that any hidden dangers of the human element are expressed and exposed.

In accident investigations, a special care is to be taken in training the investigators on these cognitive aspects. The investigation reports should have a root cause analysis that can also drill down to deeper "human error" issues rather than vague and superficial causes from which no learning or preventive actions can be formulated.

Notes & References

CHAPTER 4: HUMAN ELEMENT – THE WAY WE ARE

1. Human element: A guide to human behaviour in the shipping industry – MCA UK publication.

2. Geert Hofstede 1980 "Cultural Consequences: International Differences in Work Related Values. **Note:** Hofstede, when he was working in IBM as a psychologist between 1967 and 1973 has completed his experiment under the light of the data he collected from more than 100.000 individuals from 50 countries and 3 regions. In his famous work published in 1980 "Cultural Consequences: International Differences in Work Related Values" including the findings of this study, he considered culture as an influential power in understanding cultures and developed a four-dimensional model (Seymen 2008; Tüz and Altıntaş 2008; Clements et al. 2009). Known as Hofstede's Cultural Dimensions, these are PD, UA, I/C, M/F, and LSO (added later when the study was extended to cultures of the Far East) (Seymen 2008; Mearns and Yule 2009).

3. The role of national culture in determining safety performance: Challenges for the global oil and gas industry (K Mearns, S Yule – Safety science, 2009 – Elsevier)

4. The role of national culture in establishing an efficient safety culture in organizations: An evaluation in respect of Hofstede›s cultural dimensions (OA Seymen, OI Bolat – Balikesir University, Turkey, 2010)

CHAPTER 5: IS BBS A PANACEA?

1. Why don't people just follow the rules? A psychologist's explanation of safety management beyond behaviour based safety Dr. Rod Gutierrez Principal Psychologist DuPont Sustainable Solutions.

2. The Six Biggest Mistakes in Implementing a Behaviour-Based Safety Process: Jerry Pounds, Sr. Vice President, Aubrey Daniels International, Atlanta, USA (www.ehstoday.com) Jan. 2001.

CHAPTER 6: A PEEK INTO THE HUMAN BRAIN

1. What Are the Regions of the Brain and What Do They Do? A Nervous Journey https://askabiologist.asu.edu/brain-regions.

2. The social brain hypothesis: Robin I. M. Dunbar First published: 07 December 1998. https://doi.org/10.1002/(SICI)1520-6505(1998)6:5<178::AID-EVAN5>3.0.CO;2-8

3. Top Brain, Bottom Brain: Harnessing the Power of the Four Cognitive Modes Paperback – March 31, 2015 by Stephen Kosslyn (Author), G. Wayne Miller.

4. Schwarz, N. (2016). What anxious and angry kids need to know about their brain.

https://imperfectfamilies.com/what-anxious-and-angry-kids-need-to-know-about-their-brain/

5. Source: Boundless. "Types of Neurotransmitters by Function." Boundless Anatomy and Physiology. Boundless, 06 Jan. 2015. Retrieved 21 May. 2015.

   https://courses.lumenlearning.com/boundless-ap/chapter/neurophysiology/

6. Balancing Neurotransmitters to Take Control of Your Life Created by Deane Alban | Medically reviewed by Patrick Alban, DC. Last updated on October 17, 2017.

   https://bebrainfit.com/neurotransmitters/#the-big-four-neurotransmitters-and-how-to- balance-them

7. https://www.wsj.com/articles/how-the-brain-uses-glucose-to-fuel-self-control-1417618996

   By Robert M. Sapolsky How the Brain Uses Glucose to Fuel Self-Control.

CHAPTER 7: STRESS, ANXIETY AND FATIGUE

1. By Karen Frazier_https://stress.lovetoknow.com/How_Does_Stress_Affect_the_Brain.

2. What Anxiety Does to Your Brain and What You Can Do About It: Alan Henry.

   https://lifehacker.com/what-anxiety-actually-does-to-you-and-what-you-can-do-a-1468128356

3. https://www.healthline.com/health/chronic-fatigue-syndrome

   Medically reviewed by Stacy Sampson, DO on April 17, 2019 — Written by the Healthline Editorial Team.

CHAPTER 8: DISTORTIONS OF OUR REALITY

1. https://www.youtube.com/watch?v=lyu7v7nWzfo Anil Seth

   Your brain hallucinates your conscious reality | Anil Seth

2. 5 Ways Your Brain Plays Tricks on You By Kendra Cherry Updated March 30, 2019.

   https://www.verywellmind.com/ways-your-brain-plays-tricks-on-you-2795042.

3. https://humanhow.com/en/list-of-cognitive-biases-with-examples/The Ultimate List of Cognitive Biases: Why Humans Make Irrational Decisions: On May 30, 2017 By Tomer Hochma.

4. Simons, D. J., & Levin, D. T. (1998). Failure to detect changes to people during a real-world interaction. *Psychonomic Bulletin & Review*, 5(4), 644–649.

   http://dx.doi.org/10.3758/BF03208840

5. 15 Common Cognitive Distortions By John M. Grohol, Psy.D. Last updated: 24 Jun 2019.

   https://psychcentral.com/lib/15-common-cognitive-distortions/

6. Cognitive Distortions: When Your Brain Lies to You – Courtney Ackerman.

   https://positivepsychology.com/cognitive-distortions/

7. Maps of Bounded Rationality: A Perspective on Intuitive Judgment and Choice.

   Nobel Prize Lecture, December 8, 2002 by Daniel Kahneman.

8. Fig. 8.8 – By Fibonacci – Own work, CC BY-SA 3.0, https://commons.wikimedia.org/w/index.php?curid=17882

9. Fig 8.9 Public Domain, https://commons.wikimedia.org/w/index.php?curid=667017

10. Fig. 8.10 By Brocken Inaglory – Own work, CC BY-SA 3.0, https://commons.wikimedia.org/w/index.php?curid=8341735

11. Fig. 8.12 By Tony Philips, National Aeronautics and Space Adm. – NASA – Summer Moon Illusion (image link), Public Domain, https://commons.wikimedia.org/w/index.php?curid=1211098

12. Fig. 8.13 By Dodek – Own work, CC BY-SA 3.0, https://commons.wikimedia.org/w/index.php?curid=1529278

13. Fig. 8.14 By R. Shepard – http://mentalfloss.com/article/28862/brainworks-explaining-optical-illusions-and-other-mental-tricks, CC BY-SA 4.0, https://commons.wikimedia.org/w/index.php?curid=64969131

14. Fig. 8.15 By Jochen Burghardt – Own work, CC BY-SA 3.0, https://commons.wikimedia.org/w/index.php?curid=29475783

15. Fig. 8.16 By User: Tó campos1 – Own work, Public Domain, https://commons.wikimedia.org/w/index.php?curid=648988

CHAPTER 9: FOCUSING THE BRAIN

1. Daniel Goleman: "Focus – The hidden driver of excellence" – Bloomsbury Publishing India.

2. Exploring the Neuroscience and Magic Behind Setting Your Intent – And Creating an Optimal Future for Yourself – Kris Hallbom and Tim Hallbom

   https://www.nlpca.com/creating-an-optimal-future-for-yourself.html

3. The Intent Setting process originated from the Wealthy Mind™ program developed by Tim and Kris Hallbom in 2000.

4. What is brain fog?

   https://www.healthline.com/health/brain-fog

   Medically reviewed by Suzanne Falck, MD on June 14, 2017 — Written by Valencia Higuera

5. 10 Easy Brain Hacks To Keep You Sane At The Office

   By JOE OLIVETO Published On 07/06/2015 @Joeoliveto1

   https://www.thrillist.com/culture/brain-hacks-to-improve-mood-and-intelligence

CHAPTER 10: HOLISTIC SAFETY STRATEGY

1. "Serious Injuries & Fatalities: A Call for a New Focus on Their Prevention" (*Professional Safety*, December 2008), Fred Manuele.

2. Responsibility Allocation for Workplace Accidents.

   Ilene B. Zackowitz First Published October 1, 2001 Review Article.

   https://doi.org/10.1177/154193120104502003

3. Model Incorporating Internal and External Mechanisms to Influence Safety Behaviour. Copyright © 2010 DuPont.

   The Holistic Model for Managing Safety Copyright © 2010 DuPont.

www.ingramcontent.com/pod-product-compliance
Lightning Source LLC
Chambersburg PA
CBHW020907180526
45163CB00007B/2653